Springer Textbooks in Earth Sciences, Geography and Environment

The Springer Textbooks series publishes a broad portfolio of textbooks on Earth Sciences, Geography and Environmental Science. Springer textbooks provide comprehensive introductions as well as in-depth knowledge for advanced studies. A clear, reader-friendly layout and features such as end-of-chapter summaries, work examples, exercises, and glossaries help the reader to access the subject. Springer textbooks are essential for students, researchers and applied scientists.

Matti Leppäranta • Petteri Uotila

Geophysics of the Cryosphere

Springer

Matti Leppäranta
Institute for Atmospheric and
Earth System Research
University of Helsinki
Helsinki, Finland

Petteri Uotila
Institute for Atmospheric and
Earth System Research
University of Helsinki
Helsinki, Finland

ISSN 2510-1307 ISSN 2510-1315 (electronic)
Springer Textbooks in Earth Sciences, Geography and Environment
ISBN 978-3-031-93070-6 ISBN 978-3-031-93071-3 (eBook)
https://doi.org/10.1007/978-3-031-93071-3

This Springer imprint is published by the registered company Springer Nature Switzerland AG
The registered company address is: Gewerbestrasse 11, 6330 Cham, Switzerland

Preface

The cryosphere or the sphere of the solid phase of water forms the Earth's surface layer in high latitudes and altitudes. Its main components are atmospheric ice crystals, seasonal snow cover, frozen ground, ice sheets and glaciers, ice on lakes and rivers, and sea ice. Research of the cryosphere has been scattered across different disciplines, and, consequently, consistent overviews of the whole cryosphere are rare. Ice properties and problems are to a large degree common to the different branches of the cryosphere science, but the links can be difficult to find from the diverse literature.

Geophysics of the Cryosphere is an integrated textbook that covers the fields of this branch of science. The text begins with the physical properties of ice, introduces external forcing on the cryosphere, and then branches to the geophysics of snow, ice on land, ice in ground, ice in inland waters, and sea ice. Links between the components of the cryosphere are discussed in the text for a more complete view. The presentation is physically based that brings quantitative understanding of the phenomena and processes in the cryosphere. Also, the role of snow and ice in cold environments is discussed. The material serves as a source book for students and researchers in Earth sciences. The basis of physics and mathematics treatment is at the level of graduate students, presenting the topic from scale analysis to first-order solutions and opening views into methods and outcomes of numerical modelling.

This book has grown from the graduate course 'Geophysics of Snow and Ice' commenced in 1985 at the University of Helsinki, with both authors acting as teachers in the course of time. It has been a winter semester course with mathematical exercises and a two-day field work in boreal winter nature. The course volume is five credit points in the European Credit Transfer and Accumulation System (ECTS). The earlier books on the cryosphere have mainly a global and climate scale viewpoint or focus on the geographic description of the world's snow and ice resources, while this book is a systematic educational presentation of the geophysics of this field, serving as a course book or self-learning source material. The focus is on the physical understanding of why we have the cryosphere as it is, with its history and future predictions.

In August 2024, the United Nations General Assembly adopted the resolution to declare the period from 2025 to 2034 as the 'Decade of Action for Cryospheric Sciences'. This initiative, championed by the Republic of France and the Republic of Tajikistan and supported by a coalition of nations, aims to stimulate international scientific cooperation and foster sustainable development efforts in response to the alarming impact of climate change on the cryosphere. UNESCO has been appointed as the lead agency to coordinate this decade of action.

We are grateful to Professor Erkki Palosuo, our teacher and the pioneer in snow and ice science education in Finland, to Professors Hajo Eicken, Hardy B. Granberg, William D. Hibler III, David Holland, Knut Høyland, Elizabeth Hunke, Jouko Launiainen, Zhijun Li, Amanda Lynch, Rob Massom, Frank Nilsen, Anders Omstedt, Sergey N. Ovsienko, Yubao Qiu, Andrey Proshutinsky, Kunio Shirasawa, Ian Simmonds, Achim Stössel, Timo Vihma, and Lijuan Wen, and to many other colleagues for fruitful guidance and discussions. The course students at the University of Helsinki in the past decades are deeply thanked for their feedback in the process of course development and preparing this book. Sergey Bogdanov, Denis Demchev, Ioanna Merkouriadi, Lu Peng, Aki Vähä, and Galina Zdorovennova are deeply thanked for their careful reviews of the text. The authors also want to thank the Springer book production team for their good collaboration.

Helsinki, Finland
Helsinki, Finland
January 31, 2025

Matti Leppäranta
Petteri Uotila

Competing Interests The authors have no competing interests to declare that are relevant to the content of this manuscript.

About This Book

This book is a graduate-level textbook of the geophysics of the cryosphere or the sphere of frozen water on the Earth's high latitudes and altitudes. It is suitable for students in Earth sciences, climate research, geography, geology, glaciology, hydrology, meteorology, and oceanography. The chapter topics cover the physical properties of ice, atmospheric ice and forcing of the cryosphere, seasonal snow cover, ice sheets and glaciers, frozen ground, ice in lakes and rivers, and sea ice. The focus is on the physical understanding of why we have the cryosphere as it is, with its history and future predictions. The goal is to give a first course on the geophysics of snow and ice that dominates the nature of cold regions. This book also serves as a self-learning source book and handbook for Earth science researchers in general. The cryosphere is recognized as one of the eight branches in geophysics in the IUGG system and one of the six core projects in the World Climate Research Program.

Contents

About the Authors

Matti Leppäranta was born in 1950 in Helsinki, Finland. He received his MSc degree in mathematics in 1976 and his PhD degree in geophysics (sea ice dynamics) in 1981 from the University of Helsinki, and he performed his postdoc research at the Cold Regions Research and Engineering Laboratory (CRREL), Hanover, NH. He has worked as a research physicist at the Finnish Institute of Marine Research in 1974–1992 and as a professor of geophysics at the University of Helsinki in 1992–2018; currently, he is an emeritus professor at this university. His science career has been mostly in the geophysics of sea ice, lake ice and snow, covering their structure and morphology, remote sensing, thermodynamics, dynamics, and environmental and climate questions. He has given lectures on snow and ice in several advanced study institutes and published three books with Springer in 2004–2023: *The Drift of Sea Ice*, *Physical Oceanography of the Baltic Sea* (with Kai Myrberg), and *Freezing of Lakes and the Evolution of Their Ice Cover*. He has acted as a visiting teacher at Universitetssenteret på Svalbard (UNIS) and as a visiting professor at Hokkaido University, Dalian University of Technology, and the Aerospace Information Research Institute of the Chinese Academy of Sciences.

Petteri Uotila was born in 1969 in Tohmajärvi, Finland. He received his MSc degree in geophysics in 1994 and his PhD degree in geophysics (sea ice and physical oceanography) in 2000 from the University of Helsinki. He was a postdoc at the Courant Institute of Mathematical Sciences in New York University from 2002 to 2004. After that, he relocated to Australia and worked for Monash University, Melbourne, as a research fellow from 2004 to 2010 and then for the Commonwealth Research and Industrial Organisation (CSIRO) as a sea-ice modeller from 2010 to 2013. In 2014–2017, he was a tenure-track research professor in the Finnish Meteorological Institute, responsible for ocean climate modelling. After that, he started as a professor in geophysics at the University of Helsinki. His research focus has been on the polar oceans and their role in the climate system. He has taught and supervised oceanography students in the USA, Australia, China, Belgium, and Finland. He has written a book chapter on sea-ice modelling and on the linkages between the oceans and the rest of the climate system. He has acted as a visiting teacher at Universitetssenteret på Svalbard (UNIS), Dalian University of Technology, and the Institute of Marine and Antarctic Studies (IMAS), Tasmania.

Abbreviations

AARI	Arctic and Antarctic Research Institute (Russia)
ACC	Antarctic Circumpolar Current
AIDJEX	Arctic Ice Dynamics Joint Experiment
AIS	Antarctic ice sheet
AMSR-E	Advanced Microwave Scanning Radiometer-EOS
BP	Before present, radiocarbon dating with the reference year of 1950
BRDF	Bi-directional reflectance distribution function
CICE	Community Ice CodE (The Los Alamos National Laboratory)
CLIC	Climate and cryosphere
CMIP	Coupled Model Intercomparison Project
CRREL	Cold Regions Research and Engineering Laboratory (NH, USA)
DEM	Digital elevation model
DMSP	Defence Meteorological Satellite Program
ECMWF	European Centre of Medium-Range Weather Forecasts
ENSO	El Niño-Southern Oscillation
EOS	Earth Observing System
ERA	ECMWF Reanalysis
ERS	European Remote Sensing satellite
GIS	Greenland ice sheet
GMSLR	Global mean sea level rise
GRACE	Gravity Recovery and Climate Experiment
IACS	International Association of Cryospheric Sciences
ICESat	Ice, Cloud, and land Elevation Satellite (NASA)
IPCC	International Panel for Climate Change
IR	Infrared
IUGG	International Union of Geodesy and Geophysics
LIM	Louvain-La-Neuve
MIZ	Marginal ice zone
MODIS	Moderate Resolution Imaging Spectroradiometer
NAO	North Atlantic Oscillation
NASA	National Aeronautics and Space Administration (USA)
NCAR	National Center for Atmospheric Research (NSF, USA)
NCEP	National Centers for Environmental Prediction (MD, USA)
NDSI	Normalized Difference Snow Index
NSF	National Science Foundation

NSIDC	National Snow and Ice Data Center (Boulder, CO)
PAR	Photosynthetically active radiation
SAR	Synthetic aperture radar
SIA	Shallow ice approximation
SSA	Shallow shelf approximation
SSM/I	Special Sensor Microwave/Imager
SWE	Snow water equivalent
TIR	Thermal infrared
UNEP	The United Nations Environment Programme
USGS	United States Geological Survey
UV	Ultraviolet
WCRP	World Climate Research Programme
WMO	World Meteorological Organization

Introduction

1

Lake Kilpisjärvi, an Arctic tundra lake in the Scandinavian Mountains. *(Photograph Matti Leppäranta)*

1.1 Distribution of Snow and Ice on Earth

The global snow and ice cover or the solid phase of water forms the Earth's cryosphere.[1] Snow is a porous, soft form of ice appearing white in natural light. Snow and ice are dominant factors in the nature in cold regions, found in the atmosphere, on Earth's land and water surfaces, and in the ground (Table 1.1). The polar ice sheets in Antarctica and Greenland contain by far most of the ice volume, while seasonal snow and sea ice form the largest visible area of the cryosphere on the Earth's surface. The thickness and renewal time of the ice sheets are large, while seasonal snow cover and sea ice are thin with short renewal times. The frozen ground can be thin and seasonal or thick and perennial, and their area accounts for 35% of the Earth's land surface. In the atmosphere, the lifetime of ice crystals is of the order of days.

Cryology, the field of science covering the cryosphere, contains glaciology and has overlaps

[1] From Greek language, *'kryos' = cold*, *'sfaira' = sphere*. The words snow and ice are of Proto-Germanic origin.

M. Leppäranta, P. Uotila, *Geophysics of the Cryosphere*, Springer Textbooks in Earth Sciences, Geography and Environment, https://doi.org/10.1007/978-3-031-93071-3_1

Table 1.1 The fundamental scales of the main parts of the cryosphere and their renewal. The volume of Earth's fresh water is 35,000 × 10^3 km^2 and the surface area of Earth is 510 × 10^6 km^2. [1–3]

	Ice volume 10^3 km^3	Renewal 10^3 km^3/a	Renewal time, a	Area 10^6 km^2	Depth m
Ice sheets and glaciers	28,000	2.5	11,000	15	1900
Icebergs	7.6	1.9	4.0	N/A	N/A
Sea ice	22	23	1.0	24	0.9
Seasonal snow	2.8	5.6	0.5	24	0.1
Seasonal frozen ground	15	30	0.5	30	0.5
Permafrost	300	5.5	55	25	12
Lake and river ice	< 0.5	< 0.25	0.5	< 1	0.5
Atmospheric ice	1.7	390	0.0044	N/A	N/A

with meteorology, oceanography, hydrology, geography, geology, ecology, and engineering. Due to this multi-disciplinarity, the outcome of snow and ice research is scattered in many laboratories as well as in wide literature [4–10]. Cryology has grown into its own field of science from the scattered branches. It is presently one of the eight associations of the International Union of Geodesy and Geophysics (IUGG), named International Association of Cryospheric Sciences (IACS), and it forms one project in the World Climate Research Programme (WCRP), named Climate and Cryosphere (CliC).

In general, snow and ice occur where the air temperature of the coldest month is below the freezing point of pure water, 273.15 K (0 °C) on the Earth's surface. In the Northern Hemisphere, in January, the 0 °C isotherm is at 30°–40°N over the continents but much more north over the ocean (Fig. 1.1), while in the Southern Hemisphere, in July, this isotherm encircles Antarctica along the oceanic polar front at around 60°S. Perennial ice can be found in tundra and polar climate in ice sheets and glaciers, permafrost, and sea ice pack. In high mountains, snow and ice exist at low latitudes, and there are glaciers even at the Equator, in the Andes in Ecuador and in Mount Kilimanjaro in Tanzania.

Seasonal snow is characteristic in cold and moist climate, while lake ice and river ice extent cover the seasonal snow zone and cold and dry climate regions such as the Tibetan Plateau, where freezing surface waters exist down to 30° N, and the Antarctic Dry Valleys. Seasonal frozen ground follows largely the areas of frozen inland waters, although heavy snow accumulation can protect the ground from freezing. Sea ice forms in the polar oceans, and in the Northern Hemisphere, there are semi-enclosed freezing coastal seas down to mid-latitudes, e.g., the Baltic Sea in North Europe. Polar oceans were mapped by the great polar explorers by the early 20th century (Portrait 1.1).

1.2 Snow and Ice on Land

Snow crystals form in clouds and fall on Earth's surface through precipitation to build up the snow cover. Seasonal snow, by definition, melts in summer. Snowstorms cut down visibility and cause heavy traffic problems. Accumulation of seasonal snow cover forms a hydrological water storage which is released back to the water cycle during the melting period. Major questions in snow science are the mass balance of this storage and release of liquid melt water that necessitates good snow monitoring and melt forecasting methods [12]. Snow cover largely modifies the local heat balance and environment by its high albedo, low light transmittance, and low thermal conductivity.

Seasonal snow cover is a specific ecological environment where fine-scale snow properties play an important role. Snow provides protection from cold and predators, it is permeable for gases, and it is challenging to move in a snow-covered terrain. Snow cover has been used as a base of

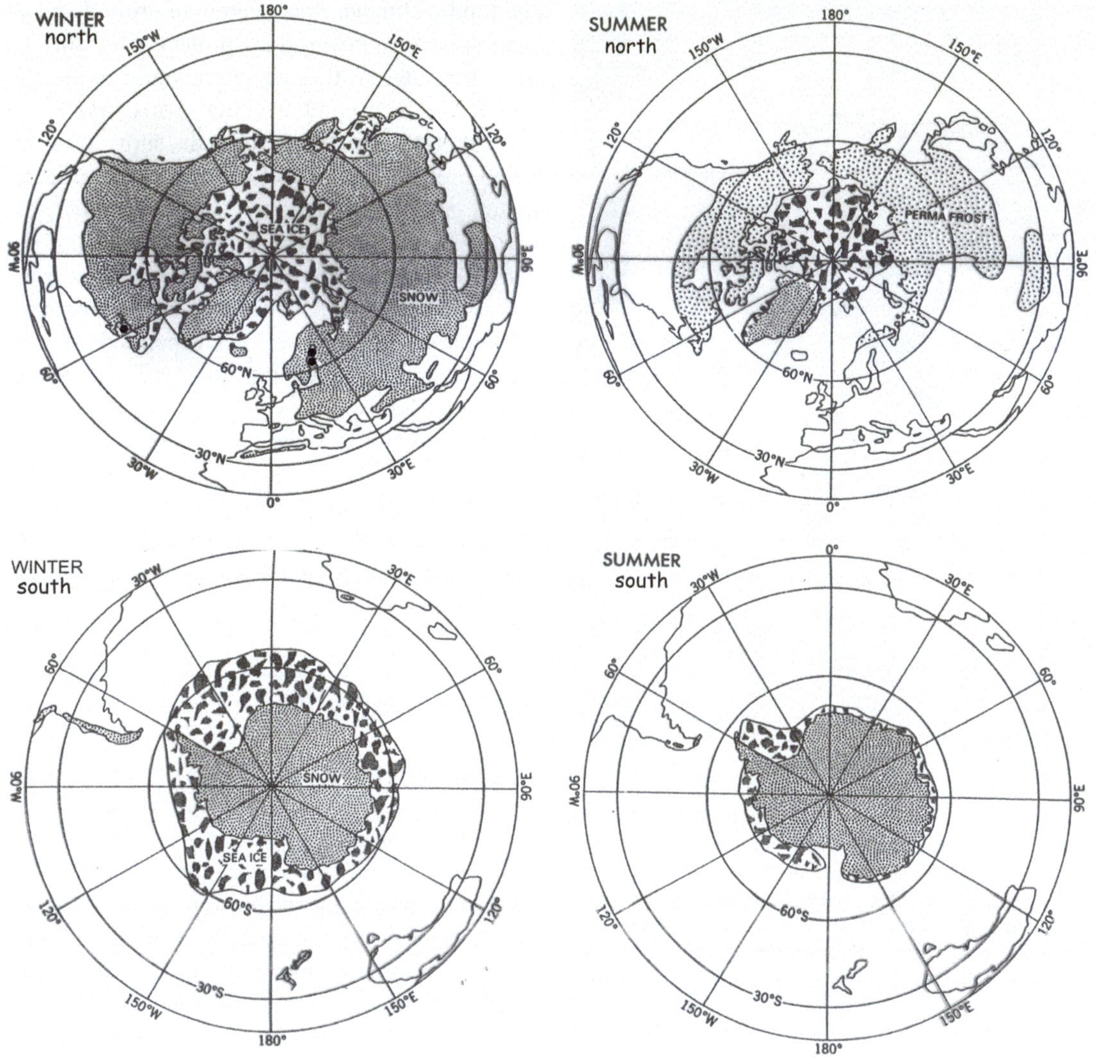

Fig. 1.1 Cryosphere in the Northern and Southern Hemisphere in winter and summer. (Drawn based on Untersteiner [3] and Huybrechts [11])

winter traffic also by people. The bearing capacity of snow undergoes evolution in snow processes, and, consequently, on-snow moving techniques change from snowshoes to skis and sledges, and further to feet and wheels [13]. In cities and highways, snow can become a nuisance since it is very expensive to remove it and transport away. Avalanches are a particular safety issue in snow-covered mountain regions. In Europe on average about one hundred people die every year in avalanches.

Where seasonal snow survives to the next winter season, new snow falls on old snow, and accumulation over many years leads to the formation of glaciers and ice sheets [14]. Two ice sheets, the Antarctic and Greenland ice sheets, exist in the present Earth, while there were others in the last Ice Age, e.g., Weichsel Ice Sheet in North Europe. The thickness of ice sheets is at most about 5 km, limited by the strength of the ice. Mountain glaciers have lateral length scales of 10–100 km and vertical scale less than 1 km. Research on ice sheets and glaciers has largely focused on their mass balances, since they regulate the mean sea level and contain large freshwater storages. Calving of icebergs is a specific

Portrait 1.1 Adolf Erik Nordenskiöld (1832–1901) was one of the Arctic research pioneers in the 1800s. He was born in Mäntsälä, Grand Duchy of Finland in the Russian Empire. He studied geology in the University of Helsinki and doctorated in 1855. Active in anti-tsarist circles for liberation of Finland, 2 years later he was forced to leave for Sweden. He acted as a professor in the Swedish Museum of Natural History in Stockholm from 1858 to the end of his life. His scientific work was focused on the Arctic, with expeditions to Greenland and Svalbard, and for which he is best known, leading the first sailing through the Northeast Passage in 1878–1879 onboard ship *Vega*. In 1893, he became a member of the Swedish Academy. (Painting by Georg von Rosen of A.E. Nordenskiöld and ship *Vega* in Northeast Passage. *Finnish Heritage Agency*)

study topic of mass balance (Fig. 1.2). Ice sheets have served as an important storage of climate information providing data of past air temperature and atmospheric geochemistry back to almost one million years ago.

In cold climate, water in the Earth's ground freezes in winter and forms the zone of frozen ground [15]. This is typically seasonal in mid-latitudes and perennial, permafrost, in polar and tundra climate. Heat content of ground and an eventual snow cover may protect land surface from freezing, so that air temperature is not the sole factor in ground freezing. For example, in parts of Hokkaido, northern Japan, snow accumulation protects the ground against freezing. If the frozen ground does not melt fully in summer, the frozen layer grows year by year and turns into permafrost which can be up to 1 km thick. Permafrost usually forms where the mean annual air temperature is below −2 °C. The permafrost zone covers about 20 percent of the land area of the Northern Hemisphere with large areas of both Eurasia and North America. Formation and melting of seasonal frozen ground are of concern for constructions and roads, while permafrost provides a stable support for buildings. Wild nature and ecology on tundra are sensitive to the freezing of soil and ground.

1.3 Floating Ice

Floating ice is a common term for ice in lakes, rivers, and seas. By Archimedes law, the ratio of freeboard to draft is about 1:10, and by thermodynamics, ice bottom temperature must be equal to the freezing point of the water just below the ice. Floating ice grows mainly from the bottom, which is limited by increasing of insulation with thickness. In one winter, floating ice does not grow more than about 2 m even in the coldest area of Siberia. Ice phenology, the bearing capacity of ice, and ice forces are the key ice questions in lakes, while dynamic phenomena, frazil ice and ice jams, are the main topics for river ice research [16]. Ice cover on lakes and rivers has an important role in the everyday life of the local population.

In marine environment, sea ice is found above the latitude 60° except in Northeast Atlantic, where the North Atlantic Current keeps the ocean open to 77°N west of Svalbard [17, 18]. In mid-latitude semi-enclosed coastal seas, such as the Gulf of St. Lawrence in Canada and the Sea of Okhotsk between

Fig. 1.2 Northern Barents Sea in summer 1980 with sea ice, a bergy bit (mid-area), and Storøya Island (80.08°N 28.05°E) covered by ice cap in the background. (*Photograph* Matti Leppäranta)

Hokkaido and Siberian east coast, sea ice occurs annually. The salinity of ocean water (35‰) lowers the freezing point to −1.9 °C and puts the temperature of maximum density even lower that has a major impact on the cooling process. In the polar seas, the drift of sea ice has an important influence on the geographic distribution and mass balance of sea ice, and the oceanic polar fronts act as boundaries which control the sea ice extent. The key questions in sea ice research are connected to cold ocean ecology, shipping in ice-covered waters, ice forces on offshore structures, interaction between sea ice and sea bottom and shore, and the role of sea ice cover in the atmosphere–ocean weather and climate systems.

1.4 People Living with Snow and Ice

Northern peoples adapted to live in snow and ice environment in prehistoric times (Fig. 1.3). Near the end of the Weichselian glaciation, about 15,000 years ago, people moved to Alaska from eastern Siberia along the Bering Land Bridge. In northern Europe, the retarding ice sheet opened good hunting and fishing grounds, and following this process, traces of human life have been found. Skis and sledges were used 9000 years BP[2] in northern Eurasia, much before the wheel was invented, and winter fishing methods were available. Snowshoes were common in northeast Asia for traveling over snow-covered terrain, and the know-how of this technique crossed the Bering Strait together with people. Inuit people adapted their life in an extreme environment in the presence of both seasonal and perennial ice [19].

Peoples in low latitudes knew of the existence of solid phase of water from high mountains and occasional snowfalls and hailstorms. In ancient times, ice was an important cooler in the hot season due to its large latent heat, collected from frozen inland waters and mountain glaciers. Ice was kept in specific stocks, such as cone-shaped ice houses called *yakhchāls* in Iran [20]. In the Bible, hail storms and snow were described several times, as phenomena and metaphors (see, e.g., Job 38), and the phases of water were understood by pre-Socrates philosophers in Greece. Pythias of Massalia (present Marseille) sailed from Gibraltar towards north in about 350–320 BC and encountered sea ice but it is not known exactly where [21], possibly in the Baltic Sea. The oldest nature book of northern Europe presents a large variety of human life in

[2] BP (Before Present) usually refers to radiocarbon dating with the reference year of 1950.

Fig. 1.3 Ice peak found in Kirkkonummi, southern Finland, dated 8500 years BP. Used for making holes in lake ice or sea ice, prepared from moose leg bone. Length 25 cm, probably used in a wooden arm. (The Finnish Heritage Agency, *Photograph* Ensio Suominen)

medieval snow and ice environment [22]: skiing, horse races and festivities on lake ice, children playing in snow, ice roads on the Baltic Sea coast, and seal hunting in marine drift ice (Fig. 1.4).

During the Little Ice Age (1450–1850), snow and ice were present in winter all over the Central Europe as illustrated in many historical paintings. Virginia Woolf described the ice winter in London with markets on the ice of river Thames in 1608 in her book *Orlando*. Knowledge about polar ice was gained in many expeditions. Willem Barentsz (1550–1597) sailed to the European sector of the Arctic Ocean in the 1590s and found Svalbard archipelago, and Adolf Erik Nordenskiöld (1832–1901) first sailed through the Northeast Sea Route in the Arctic in 1878–1880. In the south, Fabian Gottlieb von Bellingshausen (1778–1852) recorded the first observation of Antarctic continent in 1820. Attempts to reach the poles were ongoing, and the South Pole was first visited by Roald Amundsen in 1911. The conquest of the North Pole is credited to Robert Peary in 1909 but has been also claimed for Thomas Cook in 1908.

The pioneering polar explorers were followed by scientific expeditions. Karl Weyprecht led the Austro-Hungarian Expedition to Franz-Josef Land in 1872–1874 [24], and Fridtjof Nansen drifted with the *Fram* ship, moored to ice, across the Eurasian Arctic Ocean between 1893 and 1896 [25]. He was followed by the Arctic expeditions of Makarov on icebreaker *Jermak* and by Amundsen's *Maud* expedition in 1918–1925. In the south, the German ship *Deutschland* drifted with the ice in the Weddell Sea in 1911–1912 [26]. In the 1800s, empirical research was made on glaciers in the European Alps [27] and permafrost in Siberia [28]. The nature of glacier flow and permafrost became understood.

The Industrial Time has changed the human relation with the cryosphere along with traffic maintenance, snow and ice loads on structures, and recreation and tourism. Although practical snow and ice applications were utilized in prehistoric times, first scientific steps in cryosphere research were taken only in the nineteenth century. Then new technological problems arose along with railroad construction and steamships in cold regions, and general scientific interest in the Earth's nature became high (Fig. 1.5). The science of thermodynamics was established that was the key to understand the phase changes of water and evolution of snow and ice fields. Also, time series monitoring of snow and ice phenology and properties was commenced in the nineteenth century. Snow and ice possess several kinds of risks of catastrophes in modern society [29]. For example, flooding due to snow melt water drainage and frost events in summer are major issues in the agriculture of cold regions. Since the middle of the twentieth century, the development of advanced automated measurement systems, satellite remote sensing, and computation technology have brought huge steps in the development of all branches of the cryosphere science [2, 4, 7, 30].

Fig. 1.4 Carta Marina by Olaus Magnus Gothus [23] showing the Nordic countries and Northeast Atlantic. (The map was complemented by the book *The history of northern peoples* [22] with explanations for the drawings in the map)

1.5 This Book

The cryosphere is one key factor in the global climate. It controls the local and global albedo, limits the surface temperature with the freezing point as the upper boundary, and plays a role in the water cycle. The water storage in the ice sheets controls the sea level elevation on a long timescale, while seasonal snow and sea ice are renewed annually so that they may cause large year-by-year albedo variations with the positive albedo feedback phenomenon. Also, snow is a strong insulator of heat. The boundaries of the snow cover, sea ice, frozen ground, and glaciation move along with climate variations. In a warming climate, melting of permafrost releases greenhouse gases, in particular carbon dioxide and methane, thus providing positive feedback.

This book presents the geophysics of the cryosphere as a textbook of the fundamental physics. Earlier literature on cryosphere has focused on its components separately [12, 14–18] or from climate perspective (e.g., [4, 9]). After this introduction, Chap. 2 presents the Earth's hexagonal ice with its physical properties. The third chapter presents ice and snow in the atmosphere including sublimation and deposition phenomena and atmospheric forcing on the cryosphere. Then follow the seasonal snow cover, frozen ground, and land ice in their own chapters. Floating ice is treated in inland waters in Chap. 7 and in marine waters in Chap. 8, separated since the quality of ice and the spatial scales are quite different between marine and inland basins. Chap. 9 draws together the cryosphere and climate interactions. After that follows an annex with useful basic data for the

Fig. 1.5 Ilyushin IL-76D airlifter at Airport Novolazarevskaya in the Queen Maud Land, Antarctica (70°50.4′S 11°35.4′E). It has served as a stop-over for many Antarctic expeditions, from Cape Town, South Africa, to stations in the Queen Maud Land. (*Photograph* Matti Leppäranta)

study of the snow and ice on Earth. Each chapter has a list of study problems for the students.

Study Problems (the Necessary Physical Parameters Can Be Found in Annex)

1. **a** What is the fractional area of Earth's surface above latitude 60°? **b** How long does it take to make a circle along 60°S with the velocity of Antarctic Circumpolar Current, 0.1 m s^{-1}.
2. Take the Antarctic continent as a cap south of 70°S covered by an ice sheet with a mean thickness of 2.2 km. How thick water layer could form of the ice meltwater when spread evenly over the world ocean? The World Ocean covers 70.8% of the Earth's surface.
3. Seven percent of the ocean surface is covered by sea ice with an average thickness of 2 m. How is the ocean surface level affected if all sea ice melts?
4. How large a 20 cm thick river ice floe must be for you to drive on with dry feet?
5. Extreme daily net surface heat flux scales to 500 W m^{-2} in Earth's cold regions. How much this flux can (a) melt ice and (b) sublimate ice in 1 day?

References

1. Henderson P, Henderson GM (2009) The Cambridge handbook of earth science data. Cambridge University Press
2. Slaymaker O, Kelly REJ (2007) The cryosphere and the global environmental change. Blackwell
3. Untersteiner N (1984) The cryosphere. In: Houghton JT (ed) The global climate. Cambridge University Press, pp 121–140
4. Barry R, Gan TY (2011) The global cryosphere: past, present, and future. Cambridge University Press
5. Colbeck S (ed) (1980) Dynamics of snow and ice masses. Academic Press

6. Laybourn-Parry J, Tranter M, Hodson AJ (2012) The ecology of snow and ice environment. Oxford University Press
7. Marshall SJ (2011) The cryosphere. Princeton University Press
8. Pounder E (1965) Physics of ice. Pergamon Press
9. Sandells M, Flocco D (2022) Introduction to the physics of the cryosphere, 2nd edn. Institute of Physics Publishing
10. Shumskii PA (1956) Principles of structural glaciology [in Russian]. English transl. 1964. Dover Publications
11. Huybrechts P (2009) Cryosphere. In: Gornitz V (ed) Encyclopedia of paleoclimatology and ancient environments. Springer, pp 221–226Encyclopedia of Earth Sciences Series
12. DeWalle D, Rango A (2008) Principles of snow hydrology. Cambridge University Press
13. Shapiro LH, Johnson JB, Sturm M, Blaisdell GL (1997) Snow mechanics. Review of the state of knowledge and applications. CRREL Report 97-3. U.S. Army Cold Regions Research and Engineering Laboratory, Hanover, NH
14. Cuffey KM, Paterson WSB (2010) The physics of glaciers, 4th edn. Academic Press
15. Davis N (2001) Permafrost: a guide to frozen ground in transition. University of Alaska Press
16. Leppäranta M (2023) Freezing lakes and the evolution of their ice cover, 2nd edn. Springer
17. Wadhams P (2000) Ice in the ocean. Gordon & Breach Science
18. Weeks WF (2010) On sea ice. University of Alaska Press
19. Krupnik I, Aporta C, Gearheard S, Laidler GI, Kielsen-Holm L (eds) (2010) SIKU: knowing our ice. Springer, Documenting inuit sea-ice knowledge and use
20. Mahdavinejad M, Javanrudi K (2012) Assessment of ancient fridges: a sustainable method to storage ice in hot-arid climates. Asian Culture History 4(2):133–139
21. Weeks WF (1998) The history of sea ice research. In: Leppäranta M (ed) Physics of ice-covered seas, vol 1. Helsinki University Press, pp 1–24
22. Olaus Magnus Gothus (1555) Historia de gentibus septentrionalibus [A description of the northern peoples]. Olaus Magnus Gothus, Rome
23. Olaus Magnus Gothus (1539) Carta marina. Olaus Magnus Gothus, Rome
24. Weyprecht K (1879) Die Metamorphosen des Polareises [The metamorphosis of polar ice]. The Austro–Hungarian Polar Expedition of 1872–1874. Moritz Perles, Vienna
25. Nansen F (1897) Fram över Polarhavet. Aschehoug, Kristiania
26. Brennecke W (1921) Die ozeanographischen Arbeiten der Deutschen Antarktischen Expedition 1911–1912. In: Archiv der Deutschen Seewarte, XXXIX Jahrgang, Nr. 1. Hamburg
27. Clarke GKC (1987) A short history of scientific investigations on glaciers. J Glaciol 33(S1):4–24
28. Tammiksaar E, Stone IR (2007) Alexander von Middendorff and his expedition to Siberia (1842–1845). Polar Record 43(226):193–216
29. Whiteman CA (2011) Cold region hazards and risks. Wiley-Blackwell
30. Rees WG (2006) Remote sensing of snow and ice. CRC Press

Ice: Solid Phase of Water in Earth's Nature

2

I.C.E.—Indisputable Case of Emergency by AOR architects, Helsinki. This artwork consisted of nine melting ice pillars, each 13 m³, in Helsinki city centre in August 2021, symbolizing climate warming. The pillars decayed for the most in 3 days. (Photograph from news video of the Finnish broadcasting company YLE. Printed by permission)

The hexagonal ice in Earth's nature is presented, with its mechanical, thermal, and electromagnetic properties. This chapter forms the basis to the chapters that follow where snow and ice are treated in different cold environments on Earth. The last section introduces the modern key observation method in snow and ice research, remote sensing.

M. Leppäranta, P. Uotila, *Geophysics of the Cryosphere*, Springer Textbooks in Earth Sciences, Geography and Environment, https://doi.org/10.1007/978-3-031-93071-3_2

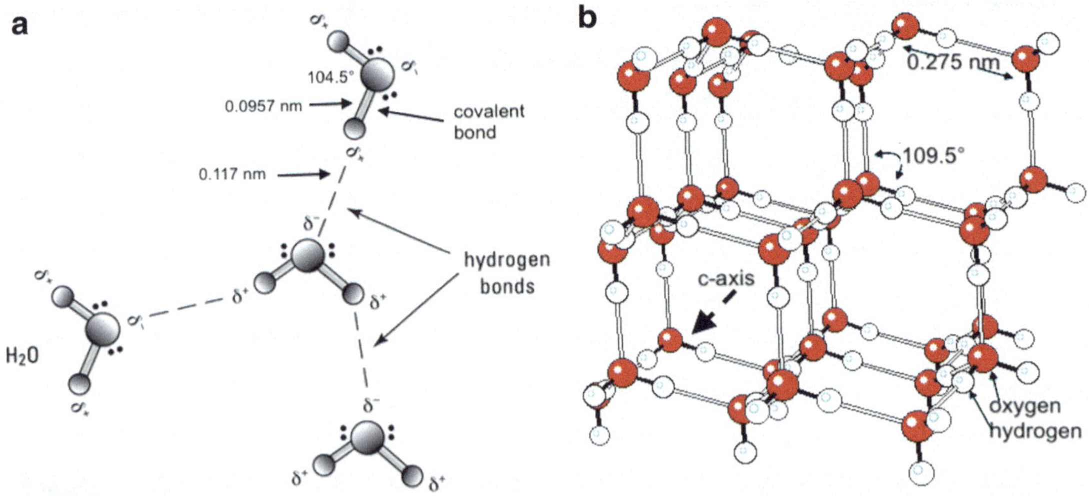

Fig. 2.1 (**a**) Water molecules in liquid state; (**b**) Ice Ih crystal lattice. The plane of the hexagons is parallel to the basal plane, and *c*-axis is normal to the basal plane (up from the paper). (Based on Petrenko & Whitworth [1] and Pounder [2])

2.1 Liquid Water and Ice

2.1.1 Water Molecule and Ice Crystal

Water is the only substance which occurs in all three phases in Earth's nature: gas, liquid, and solid. Phase transitions take place continuously. The triple point of water, where the phases are in stable equilibrium, is at (273.16 K, 611.657 Pa) [1]. This pressure is the minimum where liquid water can exist. At the Earth's surface, the pressure is much higher, from about 100 kPa at sea level down to 30 kPa at the top of Mount Everest. At the standard atmospheric pressure at sea level, p_0=101.325 kPa = 1013.25 mbar, the freezing point of water is[1] T_f = 273.15 K = 0 °C. The *freezing point depression* due to pressure is

$$\Delta T_f = -7.4 \cdot 10^{-3}(p - p_0) \tag{2.1}$$

where the pressure is in bars. The linear approximation (2.1) is good up to the pressure of 100 bar; at 1000 bar, Eq. (2.1) gives $\Delta T_f = -7.4$ °C but actually $\Delta T_f = -8.8$ °C [1].

Water molecule consists of one oxygen atom and two hydrogen atoms connected to the oxygen atom by *covalent bonds* (Fig. 2.1a), which are strong and keep the molecule stable. The hydrogen atoms are not symmetrically located that makes the water molecule an electric dipole. This structure is the cause of many exceptional properties of water, e.g., liquid water is an excellent solvent. Water vapour is invisible and less dense than dry air. In liquid water, molecules are connected by *hydrogen bonds* into linear chains where oxygen atoms are bonded with a hydrogen atom of the neighbouring water molecule. These bonds are much weaker than the covalent bonds, and they form and break continuously. The hydrogen bonds are the cause of the relatively high freezing and boiling points of water, as compared with chemically similar substances (e.g., [2]).

There are eight stable and several unstable forms of the crystal lattice structure of the solid phase of water [1]. Other than Ice I exist only in very high pressure or very low temperature. In the natural pressure and temperature within 10 km up or down from the Earth's surface, all ice belongs to the form Ice Ih, which has *hexagonal crystal structure* (Fig. 2.1b). Cubic ice, Ice Ic, occurs occasionally in the stratosphere at low pressure and temperatures below 190 K [3]. In this book, only the hexagonal ice is considered.

[1] Both Kelvin (K) and Celsius (°C) temperature scales are used in this book, whichever convenient.

The crystal lattice of Ice Ih (from now on, just 'ice') was presented by Pauling [4]. Each oxygen atom is located at a centre of a tetrahedron with four other oxygen atoms located at the apices, joined together by hydrogen bonds (Fig. 2.1b). The electron pairs are evenly spaced in their orbits, and the covalent bonds H–O–H form an angle of 109.5°, which equals the tetrahedral angle in a perfect tetrahedron. At the freezing point, the distance between neighbouring molecules is 0.275 nm. The lattice follows the Bernal– Fowler rules (e.g., [1]): (1) There are two hydrogen bonds connected to each oxygen atom, and (2) There is only one hydrogen atom in a bond. Exceptions to them are defects in the crystal lattice. The concentration of foreign molecules in ice crystal lattice is extremely small and can be ignored in Earth science research and applications.

The hexagonal hydrogen bonding is an open package so that the density of ice is less than the density of liquid water. Decrease in density at the change from liquid to solid phase is exceptional,[2] and water shows the largest decrease among all substances in Earth's nature. Thus, ice floats on liquid water. Along the *c*-axis, the crystal structure forms a hexagonal prism, where the three lateral axes are the crystal *a*-axes. The ice crystal lattice possesses hexagonal symmetry that can become visible, e.g., in the morphology of snow crystals.

Ice is optically uniaxial: light goes through undisturbed along *c*-axis, while in other directions, *birefringence* (double refraction) takes place. Molecules in the crystal lattice are concentrated in a series of parallel planes, *basal planes* (2.1b). These planes are isotropic but there is anisotropy between the basal plane and the *c*-axis direction, since there are more hydrogen bonds in the basal plane than in the planes containing *c*-axes. Thermal conductivity is highest along the basal plane that has an impact on the initial ice growth on a calm water surface: ice crystals grow laterally along the basal plane with vertical *c*-axes. Shear strength is at lowest parallel to the basal plane since then the number of bonds to be broken in shear is at minimum. Therefore, basal plane is also called slip plane or shear plane in ice mechanics.

2.1.2 Ice Nucleation

In nature, ice forms in the atmosphere, Earth's surface, and ground by deposition of water vapour or freezing of liquid water. Homogeneous and heterogeneous nucleation modes exist in the *primary nucleation* (e.g., [5]). In the former case, very strong *supercooling* (40 K) is required in pure water that can be approached in clouds. Heterogeneous nucleation takes place on particles acting as the crystallization seeds, and supercooling is much less than in homogeneous nucleation, e.g., in natural surface waters usually less than 1 K. In the *secondary nucleation*, crystals grow on existing ice surfaces. For example, snow or ice pack may undergo melt–freeze or sublimation–deposition processes. The large latent heats of freezing and deposition severely limit the volume of ice that can be produced or melted/sublimated in natural conditions.

To build an ice crystal in homogeneous fluid, Gibbs free energy ΔG is required to form the surface and the bulk particle. This energy and the minimum radius of a stable crystal, r_c, are [5], respectively,

$$\Delta G = \frac{4\pi}{3}\sigma_{cl} r_c^2,\; r_c = 2\frac{\sigma_{cl} T_f}{\rho_w L_f \Delta T_0} \quad (2.2)$$

where σ_{cl} is the surface energy of the crystal in contact with the liquid, ρ_w is the water density, L_f is the latent heat of freezing, and ΔT_0 is the amount of supercooling. For pure water, $\sigma_{cl} = 33$ mJ m^{-2} and $\Delta T_0 = 40$ K, and thus $r_c = 1.35$ nm is the critical radius of a crystal. This volume contains about 100 molecules (the diameter of a water molecule is 0.3 nm).

Heterogeneous primary nucleation takes place on suspended particles. The presence of the substrate brings the surface energies between the substrate and liquid (σ_{sl}) and substrate and crystal

[2] Bismuth, gallium, and germanium, for example, become less dense in the liquid–solid phase change. But the change is by far not so large as in water.

(σ_{sc}) into the problem (see [5]). The nucleation temperature depends on the character of the substrate, and the surface energies form a key dimensionless parameter $\cos\gamma = \frac{\sigma_{sl} - \sigma_{sc}}{\sigma_{cl}}$ where γ is taken as the angle of contact between ice and the nucleating material. Supercooling is [5]

$$\Delta T = \Delta T_0 \sqrt{\frac{1 - \cos\gamma}{2}} \qquad (2.3)$$

Ice as a substrate has $\gamma = 0$, i.e., no supercooling is needed in nucleation on ice surface. The best inorganic nucleators in nature have $\gamma = 10°$, and the supercooling would be 3.5 K, more than observed in surface waters. The difference is possibly due to that the temperature of supercooled water can vary fast in natural conditions. Field studies as well as laboratory experiments have shown that in a calm water body supercooling is 0.5–1.5 K.

2.1.3 Ice Crystal Structure

Ice core samples are examined at the site for their features including stratigraphy and visible foreign inclusions (see Portrait 2.1). Further analyses are performed in an ice laboratory (at −10 °C or so). Ice samples are first cut into *thick sections*, 1 cm thick plates, for detailed visible features, such as gas bubbles. Then *thin sections*, less than 1 mm, are processed from the thick sections, frozen on glass plates, using a plane or a microtome [6, 7]. Thin sections provide the size, shape, and the *c*-axis direction of the crystals. They are placed between crossed polarizers to see the crystals in different interference colours (Fig. 2.2). If the *c*-axis of a crystal is perpendicular to the thin section, no light goes through. That is used to find out the *c*-axis directions: the thin section is rotated in a universal stage to get the polar and azimuth angles of the *c*-axes, presented in Schmidt net[3] [7]. Crystal structure is the fundamental microscale description of ice, providing information of the properties of ice and its evolution history.

Portrait 2.1 Wilford Frank 'Willy' Weeks (1929–2017) was one of the grand ice scientists of the 1900s. He received his PhD in geochemistry in 1956 in the University of Chicago, in 1962, he joined the Cold Regions Research and Engineering Laboratory in Hanover, N.H., and in 1986, he moved to Fairbanks, Alaska where he served as Chief Scientist at the Alaskan Synthetic Aperture Radar Facility and as Professor of Geophysics in the Geophysical Institute of the University of Alaska. He greatly contributed to the research of structure and properties of ice. (His book *On sea ice* (2010) is a classic in cryosphere science. The picture shows him working with an ice sample in the Baltic Sea, March 1988. *Photograph* Matti Leppäranta)

Ice on Earth's surface is a *polycrystalline medium* consisting of ice crystal lattices and impurities, captured from the parent water or accumulated from atmospheric deposition between crystals. Dissolved matter in the parent water also depresses the temperatures of the freezing point and the maximum density T_m (e.g., [2]; Weeks 2010). In dilute solutions, these depressions are $\Delta T_f \approx -K_f S$, $\Delta T_m \approx -b_m S$, where the S is salinity in per mill (‰ or ppt), and

[3] Schmidt net is a disc illustrating the directional distribution of data in one hemisphere; *c*-axes are not vectors, so that their vertical orientations can be defined with zenith angles [0, 90°].

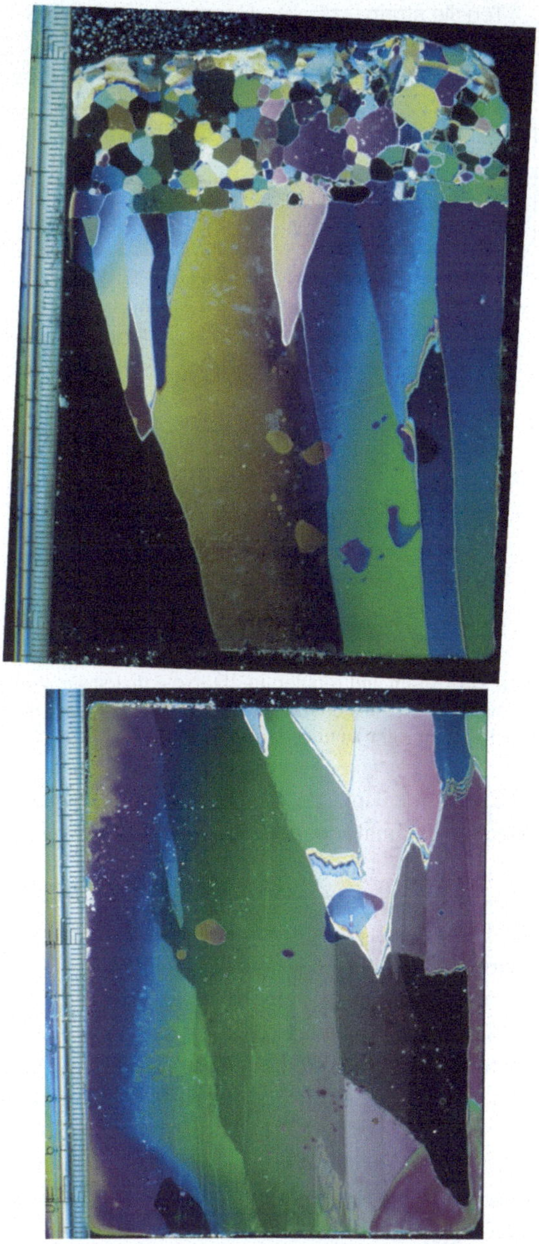

Fig. 2.2 Crystal structure of freshwater ice in Lake Pääjärvi (salinity ≈ 0.1‰), March 3, 1993 (61°03′N, 25° 10′E). The top 10.5 cm and 10.5–20 cm layers are shown. Each colour spot shows one crystal. There is a 3-cm coarse-grained (2–9 mm) snow-ice layer on top, and the lower part is columnar-grained congelation ice [8]

K_f and b_m are characteristics of the solution; e.g., for seawater, $K_f \approx -0.055$ °C and $b_m \approx -0.223$ °C.

2.2 Mechanical Properties of Ice

2.2.1 Density

At the freezing point, the densities of liquid and solid phases of pure water are 999.84 kg m^{-3} [9] and 916.7 kg m^{-3} [2], respectively. Thus, the density decreases by 8.3% at freezing. Bulk density of ice can be measured in the direct way, the mass/volume ratio of a geometrically regular sample, or by hydrostatic weighing of a sample in paraffin [10]. The density of ice has been also determined from the molecular weight of water and measured geometry of the crystal lattice within less than 0.1% from the direct measurement [2].

The density of ice increases with decreasing temperature and increasing pressure [1]. The thermal volume expansion coefficient of ice is about $1.5 \cdot 10^{-4}$ °C^{-1}, and the density of pure ice is 921.6 kg m^{-3} at −30 °C. The bulk compressibility of ice is large, about 9 GPa. For example, at 1 km depth in an ice sheet, the pressure is about 9 MPa by the hydrostatic law, so that density increases by 0.1%. The density of natural ice depends also on the gas content ν_a,

$$\rho_i(T, p, \nu_a) = (1 - \nu_a)\rho_i(T, p, 0) \tag{2.4}$$

This equation can be used the estimate the gas content of ice from a density measurement.

Example 2.1 Ice density is a fundamental property of floating ice. Denote $h = h_f + h_d$, where h is the ice thickness, h_f is the freeboard, and h_d is the draft of an ice floe. According to the Archimedes law, the weight of the floe $\rho_i h$ equals the buoyancy of the draft $\rho_w h_d$. Thus,

$$h_f = \frac{\rho_w - \rho_i}{\rho_w} h$$

In a fresh water, the freeboard of solid ice is 8.3% or $^1/_{12}$ of the ice thickness (at 0 °C). The saying that $^1/_9$ or $^1/_8$ of an iceberg is visible relates to marine environment, where the ocean water has the density of 1025 kg m^{-3}, and the density of iceberg is about 900 kg m^{-3}.

2.2.2 Ice Rheology and Strength

Mechanics of ice is examined across a wide range of length scales. Microscale concerns individual grains and ice impurities, extending from sub-millimetres to 0.1 m. Much of the field and laboratory work is done in the local scale, 0.1–10 m, where ice is a polycrystalline continuum, and the large scale extends up to the size of the ice field. In addition to polycrystalline continuum models, *broken ice fields* consisting of ice blocks are considered. They are granular media where individual ice blocks are the grains, with applications to ice jams in rivers, drift of sea ice, and avalanches.

Rheology presents the dependence of the stress on the material properties and deformation of the medium. A full 3D rheology may be mathematically quite technical and complicated (see, e.g., [11, 12]) that is avoided here. The material properties of ice have large variability due to the high homologous temperature,[4] impurities, large crystal size relative to the size of test specimens, and variations in test arrangements. In ice applications, deformation has been given by the strain and strain-rate without higher order terms. More detailed texts for ice rheology are given by, e.g., Ashton [13], Mellor [14], Cuffey & Paterson [15], Leppäranta [16], and Palmer & Croasdale [17].

In one dimension, stresses and strains are compressive or tensile. In 2D or 3D, the shape of material parcels changes due to shear strain (Fig. 2.3). In general, stresses and strains are presented as a scalar (1D) or a matrix (2&3D). The diagonal components are normal stresses or strains, and the off-diagonal components represent shear (e.g., [11, 12]). Rotation changes the orientation of a parcel but does not cause physical deformation. The fundamental rheology models are *linear elastic* or *Hooke's medium, linear viscous* or *Newton's medium*, and *ideal plastic* or *St. Venant's medium* (Fig. 2.4). In linear elastic media, stress is proportional to the strain, in linear viscous media, stress is proportional to the strain rate, and ideal plastic media fail at yield stress. Mechanical analogues of these models are spring balance for linear elasticity, dashpot[5] for linear viscosity, and static friction for plasticity. Or look to the fridge: sausage is elastic, jam is viscous, marmalade is plastic (if not, do not eat).

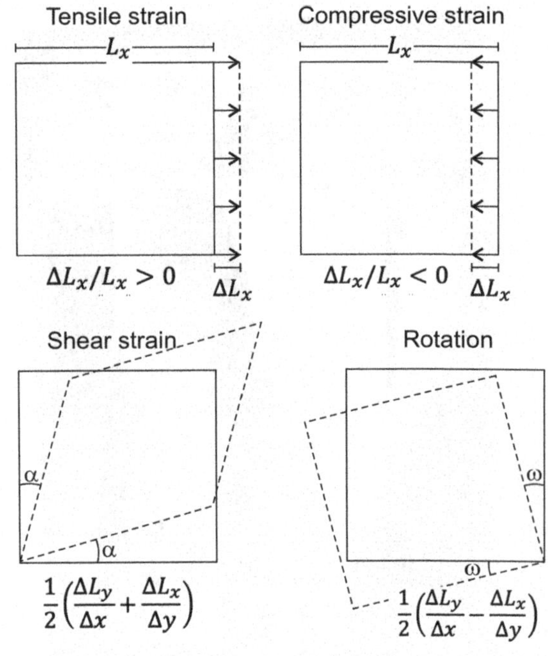

Fig. 2.3 Strain and rotation in 2D. Note that shear strain changes shape but not size, and rotation is positive anticlockwise in right-hand coordinate system

In matrix form, ice rheology is given as $\boldsymbol{\sigma}=\boldsymbol{\sigma}(\psi,\boldsymbol{\varepsilon},\dot{\boldsymbol{\varepsilon}})$, where $\boldsymbol{\sigma}$ is the stress, ψ contains material properties of the medium, $\boldsymbol{\varepsilon}$ is the strain, and $\dot{\boldsymbol{\varepsilon}}$ is the strain rate. In very short time scales (seconds) ice is a perfectly elastic–brittle material, where it fails once the strength limit has been reached. In long-time scales, ice shows non-linear viscous behaviour where the parameters have large variability depending on temperature. Plastic models have been used for a continuum approximation of the granular flow of ice blocks, but also discrete particle models have been applied for these situations.

Elastic ice. The rheology of a one-dimensional linear elastic beam reads

[4] Homologous temperature is equal to actual temperature divided by freezing point temperature (K).

[5] Such as a door stroke compressor, which dampens the movements of a door at closing.

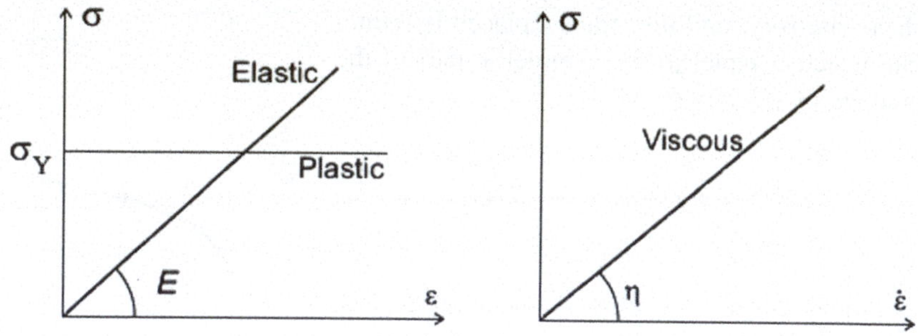

Fig. 2.4 The basic rheology models in one dimension—elastic, plastic, and viscous—for the stress σ as a function of the strain ε and strain rate $\dot{\varepsilon}$; E is the elastic modulus, η is the viscosity, and σ_Y is the yield strength

Table 2.1 Elastic properties of ice at −5 °C. The Poisson's ratio is 0.33 [13, 14]

Deformation	Modulus GPa	Brittle strength MPa
Uniaxial compression	9.0	10
Uniaxial tension	9.0	1.5
Shear	3.4	1.0

$$\sigma = E\varepsilon \tag{2.5}$$

where E is the elastic modulus or *Young's modulus* (Table 2.1). Isotropic 2D and 3D cases are analogous, stresses, and strains shown as symmetric matrixes (e.g., [11, 12]). 2D strain contains axial strains ε_{xx} and ε_{yy} and shear strains $\varepsilon_{xy} = \varepsilon_{yx}$, and the stresses are

$$\sigma_{xx} = \frac{E}{1+\mu}\left(\varepsilon_{xx} + \frac{\mu}{1-2\mu} tr\varepsilon\right),$$
$$\sigma_{yy} = \frac{E}{1+\mu}\left(\varepsilon_{yy} + \frac{\mu}{1-2\mu} tr\varepsilon\right) \tag{2.6a}$$

$$\sigma_{xy} = \frac{E}{1+\mu}\varepsilon_{xy}; \quad K = \frac{E}{1+\mu}, G = \frac{E}{2(1+\mu)} \tag{2.6b}$$

where $\mu \approx 0.33$ is the *Poisson's ratio*, defined as $\mu = -\varepsilon_{yy}/\varepsilon_{xx}$ in uniaxial tension of a beam in x-axis, tr is trace of matrix, $trA = \sum_i A_{ii}$, and K and G are the bulk modulus and shear modulus, respectively. In uniform compression, $\varepsilon_{xx} = \varepsilon_{yy}$, $\varepsilon_{xy} = 0$ and $\sigma_{xx} = \sigma_{yy} = K\varepsilon_{xx}$.

In experimental data, Young's modulus and Poisson's ratio have been within 5% from the standard values, and the *brittle strength* of ice is 1–10 MPa (10–100 kg cm^{-2}) depending on the mode of loading (Table 2.1). Compared with ice, Young's modulus and compressive strength are one order of magnitude larger for wood and two orders of magnitude larger for steel. The strength of ice increases with decreasing temperature, and the brittle compressive strength is 20 MPa at a temperature of −30 °C. In test data, at a fixed temperature and strain rate, the strength has varied by a factor of 2. The strength of porous ice has been modelled as $\sigma(\nu) \propto \sqrt{1-\nu/\nu^*}$, where ν is the porosity and $\nu^* \sim ½$ is the failure porosity (when ice breaks under its own weight).

Example 2.2 In 1D case, compression by $\sigma = 9$ MPa of a 1 m long beam causes a strain of $\varepsilon = -10^{-3}$ or the beam becomes shorter by 1 mm. Increasing the compression to 10 MPa adds 0.1 mm to the shortening and breaks the ice. Shear stress of 5 MPa causes a shear strain where the corner angles of a square change by of $0.7 \cdot 10^{-3}$ rad = 0.04°. Increasing the angle to 0.06° breaks the ice. Thus, elastic deformations are very small and hardly visible before breakage.

Speed of compressive waves is $c_1 = \sqrt{K/\rho_i} \approx 2.7$ km s^{-1}, about twice the speed of sound in liquid water, and the speed of shear waves is $c_2 = \sqrt{G/\rho_i} \approx 1.9$ km s^{-1}.

Viscous ice. In a viscous medium, stress depends on strain rate. Linear viscous rheology is analogous to the linear elastic rheology (Eqs. 2.6a, 2.6b), with the bulk modulus and shear modulus replaced by bulk viscosity and shear viscosity, and the strain replaced by strain rate which is equal to the symmetric part of the gradient of velocity (u, v):

$$\dot{\varepsilon}_{xx} = \frac{\partial u}{\partial x}, \dot{\varepsilon}_{yy} = \frac{\partial v}{\partial y}, \dot{\varepsilon}_{xy} = \dot{\varepsilon}_{yx} = \frac{1}{2}\left(\frac{\partial v}{\partial x} + \frac{\partial u}{\partial y}\right) \tag{2.7}$$

Liquid water is a viscous medium, and in natural conditions, its flow is modelled as an incompressible linear viscous fluid (e.g., [18]). Then $\nabla \cdot \boldsymbol{u} = 0$, i.e., bulk viscosity is of no concern, and shear viscosity η resists shear deformation. In simple shear flow along x-axis, $u = u(y)$, we have

$$\sigma_{xy} = 2\eta\dot{\varepsilon}_{xy} = 2\eta\left(\frac{1}{2}\frac{du}{dy}\right) = \eta\frac{du}{dy} \tag{2.8}$$

This formula is often used in illustrating viscous flow in fluid dynamics.

In ice applications, when the loading time is more than minutes and the stress is below the brittle strength, viscous response results but in a non-linear manner. The strength of viscous ice is referred as the *ductile strength*, which is less than the brittle strength and decreases with strain rate, reaching 1 MPa at the strain rate of 10^{-7} s^{-1} (0.1% strain in 3 h). Tests of viscous ice flow are usually made by monitoring the strain of the test specimen at a constant load (e.g., [13, 19]). The strain history undergoes three consecutive regimes (Fig. 2.5). Primary creep (I) is sublinear[6] or hardening ice, then a linear regime (II) follows, and the final, tertiary creep (III) is super-linear, ice fails in accelerating strain. The viscosities and time scales of the three regimes depend strongly on the ice temperature (see [20]). In the primary creep, the delayed elastic strain can be an order of magnitude larger than the immediate elastic strain, and the strain inflection point occurs at about 1% strain in the secondary creep [19]. A well-established viscous ice law, Glen's flow law $\sigma \propto \sqrt[3]{|\dot{\varepsilon}|}$, has been applied to glaciers [15, 21].

[6] Law $y(x) = cx^a$ is sublinear for $a < 1$, linear for $a = 1$, and super-linear for $a > 1$.

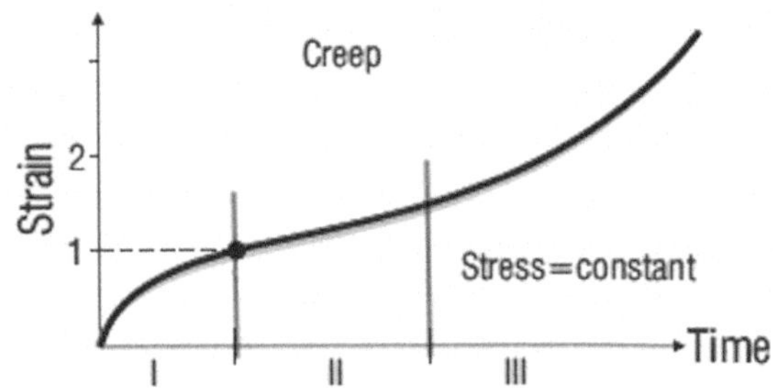

Fig. 2.5 Creep of ice under a constant load. The primary (I), secondary (II), and tertiary regimes (III) are illustrated. (Drawn based on Ashton [13])

Plastic ice. An ideal plastic medium is rigid at low stresses but fails at the yield stress σ_Y. In one dimension (Fig. 2.4),

$$\varepsilon = 0, \sigma < \sigma_Y; \quad \varepsilon \neq 0, \sigma = \sigma_Y \tag{2.9}$$

In two (three) dimensions, the yield stress is replaced by a yield curve (yield surface), which presents the yield stress for different modes of strain (e.g., [22]). At yielding, the actual strain depends on the respond of the system to forcing. Stable plastic media are strain hardening: yield strength increases under strain and more stress is needed to continue the strain. For example, when compacting a snowball, more force is needed to make it smaller, and at a fixed load, a stationary, finite snowball results. In the mechanics of systems of ice blocks, e.g., drift ice [23], yield strength is large for compression, significant for shear, and low for tension.

Generalized models. For more general models, linear systems can be changed to non-linear ones, and the basic rheological models can be combined (e.g., [14]). For example, Glen's flow law above is a non-linear viscous rheology, and linear elastic and linear viscous models in series give Maxwell medium while putting them parallel gives Kelvin–Voigt medium. For a constant load, Maxwell medium has an immediate elastic response followed by linear viscous flow, while Kelvin–Voigt medium flows in a viscous manner towards an asymptote determined by the elastic part. Viscoelastic models have been used for wave–ice interaction in sea ice research (e.g., [24]).

2.3 Thermal Properties of Ice

2.3.1 Fundamental Characteristics

Thermodynamic problems with snow and ice are due to phase changes, heat conduction, and transfer of solar energy. The latent heats of phase changes are very large (Table 2.2; Fig. 2.6), and therefore daily changes in ice volume are necessarily small in natural conditions. Loss of ice per unit surface area in time Δt is

$$\Delta h = \frac{Q}{\rho_i L_p} \Delta t \tag{2.10}$$

where Q is the heat flux, ρ_i is the ice density, and L_p is the latent heat of phase change. For a large net heat flux in Earth's cold regions, $Q = 100$ W m^{-2}, we have $\Delta h = 2.8$ cm for ice melting and $\Delta h = 3.3$ mm for sublimation in 1 day.

In many applications, fixed values are used for the thermal properties of ice and water. Thermal diffusion coefficient

Table 2.2 Thermal properties of pure ice and liquid water

Property	Symbol	Unit	Ice 0 °C	Ice −20 °C	Liquid water 0 °C
Density	ρ	kg m^{-3}	916.7	920.3	999.84
Latent heat of melting	L_f	kJ kg^{-1}	333.5	–	–
Latent heat of sublimation	L_e	MJ kg^{-1}	2.82	2.84	–
Latent heat of evaporation	L_e	MJ kg^{-1}	–	–	2.49
Specific heat	c	kJ kg^{-1} °C^{-1}	2.11	1.96	4.22
Thermal conductivity	k	W m^{-1} °C^{-1}	2.14	1.88	0.561
Thermal diffusion coefficient	D	m^2 s^{-1}	$1.11 \cdot 10^{-6}$	$1.04 \cdot 10^{-6}$	$0.133 \cdot 10^{-6}$

Fig. 2.6 Since ancient times until mid-1900s, lake ice was sampled in winter to be used as a cooler in the warm season. The picture shows ice harvesting in Lake Tuusulanjärvi, Finland in the 1960s. (*Photograph* by Eino Nikkilä. Reproduced with permission from the Cultural Society of Järvenpää Community)

Table 2.3 Thermal properties of ice, snow, iron, pinewood, air and water

	Density kg m^{-3}	Specific heat kJ kg^{-1} °C^{-1}	Conductivity W m^{-1} °C^{-1}	Diffusion coefficient 10^{-6} m^2 s^{-1}
Ice (0 °C)	917	2.11	2.14	1.1
Seasonal snow (0 °C)	100–300	2.11	0.029–0.23	0.14–0.36
Iron	7800	0.44	76	22
Pinewood	450	1.2	0.2	0.4
Air (0 °C)	1.23	1.005	0.0244	20
Water (0 °C)	999.84	4.22	0.561	0.13

$$D = \frac{k}{\rho c} \tag{2.11}$$

contains the time $\mathcal{T}$ and length $\mathcal{L}$ scales of diffusive heat flow with $\dim D = \mathcal{L}^2\mathcal{T}^{-1}$. For example, the diffusion length scale $\sqrt{D\mathcal{T}}$ in ice is 31 cm for daily and 5.9 m for annual cycle. In comparison between ice and other substances (Table 2.3), heat diffusion is much faster in iron but much lower in pinewood. In liquid water, specific heat is higher, and conductivity is lower than in ice (Table 2.2). In marine and inland waters, the flow is in turbulent state, and the turbulent conductivity is several orders of magnitude larger than the molecular value. Ice cover is an effective insulator in a water body because it replaces turbulent heat transfer by molecular conduction in the solid surface layer. A snow layer on ice further reduces the heat flux by a very low heat diffusion coefficient (Table 2.3).

2.3.2 Heat Conduction in Snow and Ice

Heat is conducted from warm to cold. Since the temperature of snow and ice is never above the melting point, conduction cannot melt them inside but only from boundary. According to the Fourier heat conduction law, the heat flux is proportional to the temperature gradient:

$$Q = -k\nabla T \tag{2.12}$$

where k is the thermal conductivity. The heat content of a unit volume, $\mathcal{H} = \rho c T$, changes at the rate of $\nabla \cdot Q$ by conduction. Absorption of solar radiation only can act as an internal heat source in natural conditions.

Heat flows primarily in the vertical direction in snow and ice fields, and the diffusion length scale is small. Therefore, one-dimensional approach is usually taken for the heat conduction law:

$$\frac{\partial}{\partial t}(\rho c T) = \frac{\partial}{\partial z}\left(k\frac{\partial T}{\partial z} + Q_s\right) \tag{2.13}$$

where Q_s is the solar radiation, and z is the vertical coordinate, zero at the surface and positive upward. Fixed (Dirichlet), flux (Neumann), or mixed boundary conditions can be used. The classical solution is based on constant thermal properties without the source term in a semi-infinite space. The boundary conditions are a temperature sine wave at the surface and a fixed deep value. Thus,

$$\frac{\partial T}{\partial t} = D\frac{\partial^2 T}{\partial z^2} \tag{2.14a}$$

$$z = 0 : T(0,t) = \overline{T} + \Delta T \sin(2\pi\omega t + \varphi) \tag{2.14b}$$

$$z \to -\infty : T(z,t) \to \overline{T} \tag{2.14c}$$

where $\overline{T}$ and ΔT are the mean and amplitude of the surface temperature, ω is the frequency, and φ is the phase. This system is useful in many problems when the depth is $d \gg \sqrt{D/\omega}$. The solution is:

$$T(z,t) = \overline{T} + \Delta T e^{\lambda z}\sin(2\pi\omega t + \varphi + \lambda z);$$
$$\lambda = \sqrt{\frac{\pi\omega}{D}}, z \leq 0 \tag{2.15}$$

The parameter λ contains the information of the medium and forcing frequency. Its inverse equals the *e*-folding depth of the temperature amplitude, and at depth $\pi\lambda^{-1}$ the phase lag is half cycle, and the amplitude has damped to 4%. The solution (2.15) can be used when solar radiation is ignored, or when the diffusion length scale is much larger than the optical depth.

Example 2.3 For daily cycle, $\omega = \mathrm{d}^{-1}$; $\lambda^{-1} \approx 17$ cm and the depth of 12-h phase lag is 54 cm; for annual cycle, $\omega = \mathrm{a}^{-1}$; $\lambda^{-1} \approx 3.3$ m and the depth of 6-month phase lag is 10.4 m.

2.3.3 Ice Deterioration

Ice deterioration is a coupled thermal and mechanical process. Heat is required for phase changes, and mechanical breakage of weakened melting ice speeds up the loss of ice by adding exposed surface area. Boundary fluxes melt at the boundary, and solar radiation penetrating to ice melts it inside that also weakens the ice. For example, icebergs decay fast, when they are broken to smaller pieces by wave action in open ocean.

Surface melting takes place, when the surface temperature is at the melting point, and net heat flux is positive

$$\rho L_f \frac{dh}{dt} = -k\frac{\partial T}{\partial z}\bigg|_{z=0^-} + Q_{0^+} > 0 \qquad (2.16)$$

where Q_{0^+} is the net surface heat flux from above and dh/dt is the rate of melting. Melting at the bottom is modelled in corresponding manner. The top surface may also lose snow and ice by sublimation (see Sect. 3.3). Internal melting is treated with the porosity ν which increases due to solar radiation and decreases due to heat conduction:

$$\frac{\partial \nu}{\partial t} = \frac{1}{\rho L_f}\left(k\frac{\partial^2 T}{\partial z^2} + \frac{\partial Q_s}{\partial z}\right) \qquad (2.17)$$

where $\nu \geq 0$ and $T = T_f$. If a radiation flux of 35 W m^{-2} is absorbed in a 1-m ice layer, the average porosity increases by 1% per day. The contribution of internal melting depends on the weather and optical properties of the ice.

The difference between conductive heat loss and absorption of solar radiation is at maximum below the surface at depth depending on the optical depth of the medium. For example, in the boreal zone, melting of snow is often observed in spring first at about 10 cm beneath the surface. In a blue ice spot in Antarctic ice sheet, ice melting begins half metre below the surface due to solar radiation, when the surface temperature is still several degrees below the melting point [25]. Thus, transparent snow or ice surface acts as one kind of a greenhouse ceiling.

2.4 Electromagnetic Radiation Transfer in Snow and Ice

2.4.1 Electromagnetic Radiation Spectrum

Electromagnetic radiation spectrum covers a wide of range wavelengths, from gamma radiation to radio waves (Table 2.4). In snow and ice applications (e.g., Rees [26]), the related research topics concern measurement technology (gamma rays and X-rays), solar radiation (ultraviolet to short-wave infrared), and terrestrial radiation (thermal infrared to microwaves). Remote sensing applications utilize the whole spectrum. The spectrum is looked in the wavelength (λ) or frequency (f) space, $c_0 = \lambda f$ where c_0 is the velocity of light in vacuum.

Radiation transfer is described using radiance and irradiance (Fig. 2.7). *Radiance*, $L = L(r, \lambda)$ is the radiant flux emitted by an infinitesimal element of the source area in a direction r through an infinitesimal element of solid angle[7] (e.g., [26–28]). The unit of radiance is W m^{-2} nm^{-1} sr^{-1}. *Irradiance*, $E(z, \lambda)$ is the incident radiance on an infinitesimal element integrated over a hemisphere (unit W m^{-2} nm^{-1}). Downwelling and upwelling planar irradiances at a vertical level z (positive up) are given by, respectively,

[7] Solid angle of a cone is its area on a unit sphere, given in steradian (sr). For full sphere, it is 4π.

Table 2.4 Wavelength bands of electromagnetic radiation. The band limits show minor variability in literature

Window	Wavelengths (frequencies)	Sub-bands	
Gamma rays	1–10 pm		
X-rays	0.01–10 nm		
Ultraviolet	100–400 nm	UV-C	100–280 nm
		UV-B	280–315 nm
		UV-A	315–400 nm
Optical or light [photosynthetically active radiation (PAR)]	380–760 nm [400–700 nm]	Violet	380–450 nm
		Blue	450–495 nm
		Green	495–570 nm
		Yellow	570–590 nm
		Orange	590–620 nm
		Red	620–760 nm
Infrared	760 nm—1 mm	Near IR (IR-A)	760–1400 nm
		Short-wave (IR-B)	1.4–3 μm
		Thermal (IR-C)	3–30 μm
		Far IR	30–1000 μm
Microwaves	1 mm –1 m (0.3–300 GHz)	Millimetre waves	1–10 mm
		Radar	1 cm—1 m
Radio waves	> 1 m (< 300 MHz)	Short	1–100 m
		Medium	0.1–1 km
		Long	> 1 km

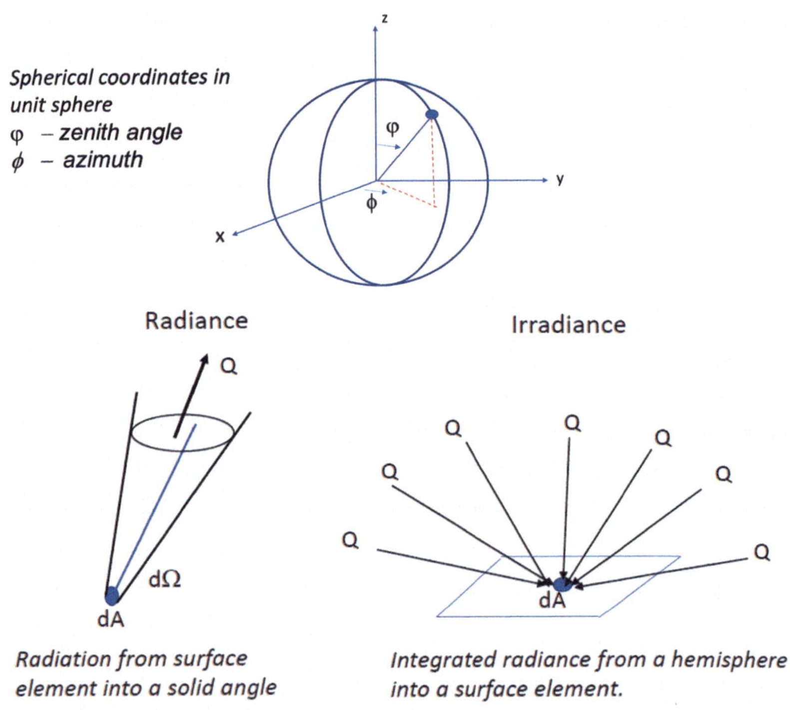

Fig. 2.7 Spherical coordinate system on a unit sphere with zenith angle φ ($0 \le \varphi \le \pi$), and azimuth ϕ ($0 \le \phi \le 2\pi$). Below a schematic picture of the basic concept of radiation: radiance and irradiance, both have the dimension of (radiation) power per unit area

$$E_d(z,\lambda) = \int_0^{2\pi} d\phi \int_0^{\pi/2} L(z,\varphi,\phi,\lambda)\cos\phi\sin\varphi d\varphi \tag{2.18a}$$

$$E_u(z,\lambda) = \int_0^{2\pi} d\phi \int_{\pi/2}^{\pi} L(z,\varphi,\phi,\lambda)\cos\phi\sin\varphi d\varphi \tag{2.18b}$$

where φ ($0 \leq \varphi \leq \pi$) is the zenith angle, and ϕ ($0 \leq \phi \leq 2\pi$) is the azimuth (e.g., [28]). Integrating the planar irradiances with respect to the wavelength, the downwelling and upwelling energy fluxes are obtained (unit W m^{-2}).

The physical basis of energy transfer by radiation is the *Planck's law of black body radiation*: the emitted electromagnetic radiance is a function of temperature,

$$L_B(\lambda;T) = \frac{2hc_0^2}{\lambda^5\left[\exp\left(\frac{hc_0}{\lambda k_B T}\right) - 1\right]} \tag{2.19}$$

where h is the Planck's constant, and k_B is the Boltzmann's constant. This function vanishes at $\lambda \to 0$ and $\lambda \to \infty$ and has a single peak at a fixed temperature. With increasing temperature, the level of radiation increases, and the peak shifts towards short wavelengths. Integrating Planck's law over the wavelength and solid angle, the emitted radiative energy flux is given by the *Stefan–Boltzmann law*

$$Q_B = \sigma T_0^4 \tag{2.20}$$

where σ is the Stefan–Boltzmann constant, and T_0 is the surface temperature. At the freezing point, black body radiation is 316 W m^{-2}. The grey body model assumes Planck's spectrum reduced everywhere by a fixed emissivity coefficient ε_0, $0 < \varepsilon_0 < 1$; thus, its emission is $\varepsilon_0 \sigma T_0^4$. This model is used for snow, ice, and water surfaces with $0.96 < \varepsilon_0 < 0.98$.

2.4.2 Optical Properties of Snow and Ice

Radiative transfer in snow and ice takes place mostly in the optical band or sunlight (380–760 nm) which contains approximately half of the solar energy. The penetration depth is 0.1–10 m for ice and 5–10 cm for snow. Shorter and longer wavelengths are absorbed in a near-surface layer with thickness less than about 10 cm. A common approach in snow and ice research is to divide solar radiation into the sunlight, which acts as the source term in the heat conduction equation (Eq. 2.13), and the rest (ultraviolet and infrared bands) is added into the surface boundary condition.

Optical properties of transparent media are classified to *inherent* and *apparent optical properties* [27, 28]. Inherent optical properties depend on the medium only and are based on the absorption coefficient and the scattering function. *Absorption* follows the Beer–Lambert law:

$$\frac{dL}{dr} = -aL \tag{2.21}$$

where r is the coordinate in the direction of light propagation, and $a = a(r, \lambda)$ is the beam absorption coefficient. The solution is

$$L(r,\lambda) = L(0,\lambda)\exp\left[-\int_0^r a(s,\lambda)ds\right] \tag{2.22}$$

If a does not depend on r, we have $L(r, \lambda) = L(0, \lambda)e^{-a(\lambda)r}$. Light absorption spectrum of pure ice is much as that of pure liquid water (Fig. 2.8). Due to hydrogen bonding, peaks in the ice absorption spectrum are shifted towards lower energy. Ultraviolet and infrared fractions of solar radiation attenuate fast in pure ice and water, for wavelengths longer than 1200 nm the absorption coefficient is more than 1 cm^{-1} (see [29]).

The *scattering function* is $\beta = \beta(\lambda, \psi)$, where ψ is the scattering angle, $0 \leq \psi \leq \pi$ (e.g., [27, 28]). The scattering function is usually integrated for the forward scattering and backscattering coefficients b_f and b_b, respectively. Scattering diffuses radiance and therefore reduces the radiance level in the direction of propagation, as specified by the total scattering coefficient $b = b_f + b_b$. The beam attenuation law for scattering is similar with the Beer–Lambert law, with a replaced by b in Eq. (2.21). Finally, assuming additivity of absorption and scattering contributions, we have

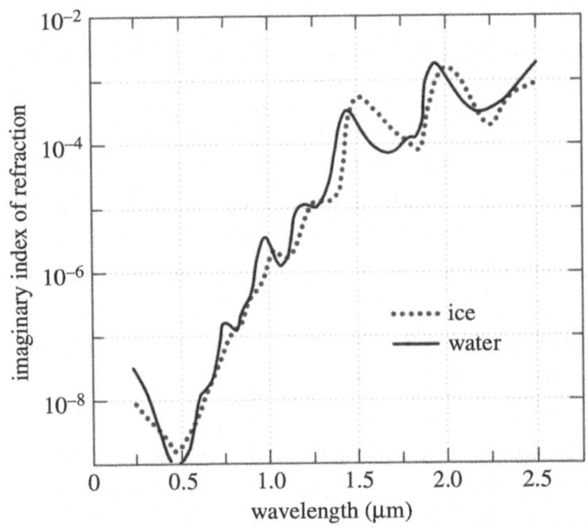

Fig. 2.8 Imaginary index of refraction m_{Im} of ice and water. Absorption coefficient is $a = 4\pi m_{Im}/\lambda$,in water $a(0.5\ \mu\text{m}) = 0.025\ \text{m}^{-1}$, $a(0.7\ \mu\text{m}) = 0.65\ \text{m}^{-1}$. (*From* open-source article by Warren [29], *Philosophical Transaction of the Royal Society A*)

$$\frac{dL}{ds} = -cL \tag{2.23}$$

where $c = a + b$ is the *beam attenuation coefficient*.

Apparent optical properties are connected to irradiance, and they depend on the directional distribution of the incident radiance in addition to the properties of the medium. *Spectral albedo* or *spectral reflectance* α_λ and the (total) albedo α are obtained as the ratio between upwelling irradiance and downwelling irradiance just above the surface:

$$\alpha_\lambda = \frac{E_u(0^+,\lambda)}{E_d(0^+,\lambda)}, \alpha = \frac{\int E_u(0^+,\lambda)d\lambda}{\int E_d(0^+,\lambda)d\lambda} \tag{2.24}$$

Upwelling irradiance consists of reflection from the surface and scattering from below. Snow and ice are highly scattering media, while liquid water albedo is much lower, and reflection dominates in the upwelling irradiance.

In the first-order approximation, attenuation of irradiance in snow and ice is modelled as an analogue to the beam attenuation law (Eq. 2.23):

$$\frac{d}{dz}E_d(z,\lambda) = \kappa_\lambda(z)E_d(z,\lambda) \tag{2.25}$$

where κ_λ is the *(diffuse) attenuation coefficient*, also called the 'extinction coefficient' (cf. beam attenuation coefficient in Eq. 2.23). This equation integrates into

$$E_d(z,\lambda) = E_d(0^-,\lambda)\exp\left[-\int_z^0 \kappa_\lambda(\zeta)d\zeta\right], z \leq 0 \tag{2.26}$$

When the attenuation coefficient is independent of depth, the exponent is $\kappa_\lambda z$, and κ_λ^{-1} is the spectral *light penetration depth*. At the depth of $3\kappa_\lambda^{-1}$, irradiance is attenuated to 5% of the surface level. The *transmittance* $\tau_\lambda(h)$ and *absorbance* $A_\lambda(h)$ of the layer $[-z, 0]$ are

$$\tau_\lambda(z) = \frac{E_d(z,\lambda)}{E_d(0^+,\lambda)}, A_\lambda(h) = \frac{E_d(0^-,\lambda) - E_d(z,\lambda)}{E_d(0^+,\lambda)} \tag{2.27}$$

Clearly $\alpha_\lambda + A_\lambda + \tau_\lambda = 1$.

Integrating Eq. (2.26) over the wavelength gives the absorption of solar heat with respect to depth, which is the source term in the heat conduction law (Eq. 2.13):

$$\frac{\partial Q_s(z)}{\partial z} = \frac{\partial}{\partial z}\int_0^\infty E_d(z,\lambda)d\lambda = \int_0^\infty \kappa_\lambda(z)E_d(z,\lambda)d\lambda \tag{2.28}$$

If $\kappa_\lambda = \kappa$ is spectrally constant, the last term is $\kappa Q_s(z)$. This is a good approximation for dry snow, but the attenuation spectrum of clear ice has a strong minimum in blue light that has significant impact on light transfer.

The main optical impurities of snow and ice are gas and liquid water. Gas bubbles increase scattering and reduce light penetration, while liquid water pockets act in opposite manner. Since radius of gas bubbles is much greater than the wavelength of light, the scattering spectrum is even. Absorption of sunlight is mostly used for heating, and, to less degree, for chemical energy and photosynthesis. Absorption increases with wetness, which leads to an important consequence: positive feedback between meltwater and albedo. We can say that liquid water is hostile to snow and ice environment.

Photosynthetically active radiation (PAR) covers the band [$\lambda_1 = 400$ nm, $\lambda_2 = 700$ nm]. Primary production takes photons from all directions, which is included in *scalar irradiance*, the irradiance on an infinitesimal sphere, given (cf. planar irradiances in Eqs. 2.18a, 2.18b) by

$$E_s(z,\lambda) = \int_0^{2\pi} d\phi \int_{-\pi/2}^{\pi/2} L(z,\varphi,\phi,\lambda)\sin\varphi d\varphi \tag{2.29}$$

The energy of a photon is hc_0/λ, and the number of photons or light quanta is expressed in moles.[8] Their flux, *the quantum irradiance*, is obtained by integrating the irradiance for the photon energy (e.g., [28]):

$$P(z) = \int_{\lambda_1}^{\lambda_2} \frac{\lambda}{hc_0} E_s(z;\lambda) d\lambda \tag{2.30}$$

Usually, quantum irradiance is expressed in μmol m^{-2} s^{-1}. The solar quantum flux is 10^3–$5{\cdot}10^3$ μmol m^{-2} s^{-1} at the Earth's surface, and the level of quantum irradiance needed to maintain primary production is of the order of 25 μmol m^{-2} s^{-1} in natural waters, which corresponds to irradiance power of around 5 W m^{-2}.

The colour of snow and ice. A human eye sees the colour of snow and ice from the reflection and scattering of light. Reflection comes from the surface with an even spectrum, while scattering brings in the colour of the medium (see [30]). Thick, dry, and pure snow cover is white due to scattering. The white colour can be contaminated by liquid water of wet snow, dirt, such as black carbon and dust, and snow algae. Thick, pure ice is blue (see the absorption spectrum, Fig. 2.7) because long wavelengths are largely absorbed when the path lengths inside the ice becomes long. The natural blue colour is seen only in very thick and solid ice such as bare and clean glacier spots and icebergs.

2.5 Remote Sensing

2.5.1 Methodology

Remote sensing is the class of methods to detect the characteristics of a target without direct contact. Usually this is based on a satellite, aircraft, or a fixed platform with sensors for receiving reflected and scattered or emitted electromagnetic radiation from the target. Active methods, such as radars, send signals and analyse the target from the return. Satellite methods have wide lateral coverage with data up to 50 years ago. The lateral properties of snow and ice are usually well observed since their signals contrast well with environment, but information of the vertical extent is limited. Airborne platforms are feasible for methods such as radio echo-sounding of glaciers. Acoustic method provides another type of remote sensing, though limited in practice. The signal can travel long distances in ice and liquid water. E.g., attempts have been made to use sound waves to measure the thickness of a lake ice cover [31].

Remote sensing has become a highly effective and utilized tool in mapping snow and ice extent, properties, and processes (e.g., [26, 32]). Ground-based photography has documented interesting snow and ice phenomena since mid-1800s. Airborne remote sensing was commenced in the early 1900s. A new view opened over of snow and ice fields that meant great progress, e.g., to understand the morphology and drift of sea ice fields. Recently, drone systems have provided a feasible and flexible method for local studies, with many types of sensors. Since the 1980s, satellite methods have progressed with huge steps (Table 2.5). The temporal resolution depends on the sensor, satellite orbit system, and swath width. For example, the MODIS (Moderate Resolution Imaging Spectroradiometer) and AMSR-E (Advanced Microwave Scanning Radiometer—Earth observing system)

[8] The number of quanta in one mole equals the Avogadro's number $N_A = 6.022142{\cdot}10^{23}$.

Table 2.5 Satellite windows in remote sensing

Window	Observation	Snow and ice information	Limitations	Examples
Optical	Colours	Snow and ice extent,	Daylight	Landsat
Near-infrared	False colours	Ice type, fractures	Clouds	Aqua/Terra MODIS Sentinel-2
Thermal infrared	Surface temperature	Snow and ice extent	Clouds	
Microwaves	Brightness temperature Backscatter	Surface temperature Surface layer quality	None	DMSP SSM/I Aqua AMSR-E Sentinel-1

MODIS Moderate Resolution Imaging Spectroradiometer, *DMSP* Defence Meteorological Satellite Programme, *SSM/I* Special Sensor Microwave/Imager, *AMSR-E* Advanced Microwave Scanning Radiometer—Earth observation system

Fig. 2.9 Copernicus Sentinel-3 image over North Europe on March 6, 2017. Scandinavian mountains have seasonal snow, drift ice exist in the northern Baltic Sea and White Sea, small lakes have ice cover but a large one, Ladoga in the east has drift ice. (*Source* European Space Agency)

instruments of Aqua/Terra satellite provide daily moderate (250–1000 m) and low-resolution (5–50 km) data, respectively, depending on the wavelength, but the high-resolution (15–30 m) Landsat has a repeat cycle of 16 days. Satellite sensors measure returned solar radiation, radiation emitted by Earth surface, or return of own laser or radar signal (Fig. 2.9).

Example 2.4 The period (t) of satellite orbits depends on the distance from the centre of Earth (R), Earth's mass (M), and the gravitational constant (G). By dimensional analysis, it can be shown that $\frac{t^2 MG}{R^3}$ = constant, and celestial mechanics tells that constant $= 4\pi^2$. E.g., the polar orbiting Aqua/Terra satellites are at distance 705 km from the Earth's surface; $R = 6371$ km + 705 km, and thus $t = 99$ min.

2.5.2 Optical and Thermal Remote Sensing

Optical and near-infrared sensors measure the solar radiance returning from Earth's surface (Table 2.5). The recorded radiance depends on the physical properties of the surface layer of the target, solar altitude, atmospheric conditions, and the sensor technique. The key properties of snow and ice are the surface roughness, size and shape of ice crystals, gas and liquid water inclusions, and the content other impurities [29, 32]. Snow and ice surfaces are often assumed Lambertian, i.e., have a perfectly diffuse reflectance. This assumption is not always justified, and therefore bi-directional reflectance distribution function (BRDF) models have been employed (e.g., [33]). The data can be used for snow and ice extent, surface features such as fractures, and ice types.

Example 2.5 The normalized difference snow index (NDSI) has been widely used to identify snow pixels [34]. For MODIS data, the expression is.

$$\text{NDSI} = \frac{\text{band4} - \text{band6}}{\text{band4} + \text{band6}}$$

where band4 and band6 are the reflectance of the wavelength bands 545–565 nm (green) and 1628–1652 nm (short-wave infrared), respectively. Snow pixels are assumed to have $NDSI > 0$. Snow has very high reflectance in the optical window and very low reflectance in the short-wave infrared, a characteristic used to distinguish snow from most cloud types.

Thermal infrared (TIR) remote sensing technique reveals surface objects with different temperature or emissivity [35]. Thermal radiation follows the law $\varepsilon(\lambda)L_B(\lambda, T_R)$, where ε is the emissivity, and L_B is the black body radiation (Eq. 2.19). This law peaks near 10 μm at Earth's surface temperature, and $\varepsilon = 0.998$ (snow), 0.997 (ice), and 0.995 (water) [36]. Measurements are made near the peak wavelength, and the surface temperature is obtained from the radiant temperature T_R as

$$T_0 = \frac{1}{\varepsilon} T_R \tag{2.31}$$

The snow and ice radiant temperature is thus lower than the real temperature by 0.5–1 °C. Clear sky conditions are needed for satellite measurements. The signal comes from a millimetre-scale surface layer but contains information of thermally active surface layer via the thermal properties of the medium (e.g., [37]). This method is also used for monitoring of the location of the melting front in spring.

2.5.3 Microwave Remote Sensing

Microwave window is very useful for snow and ice targets since atmosphere is practically transparent for several microwave channels (e.g., [26, 38]). *Microwave radiometry*, or passive microwave method, records the thermal radiation of Earth's surface as does the TIR method but uses microwave channels, and *radar* (radio detecting and ranging), or active microwave method, has its own signal for the measurements. Transmitted and received signals have horizontal or vertical polarization that influences the information obtained from the target. In satellite remote sensing, the passive method has provided daily global low-resolution data since 1978, while spatially and temporally limited high-resolution synthetic aperture radar (SAR) data are available from 1992.

Microwave emissivity is much lower for liquid water than for ice. Therefore, both emissivity and temperature can guide the distinction between snow or ice and liquid water. The measured signal comes from surface layer and provides information of the existence and quality of snow and ice. The emitted radiation is expressed as the *brightness temperature* T_B, which is equal to the temperature of a black body that would provide the same radiance:

$$L_B(\lambda; T_B) = L(\lambda; T) \tag{2.34}$$

where L_B and L are the radiative spectra of black body and the target, respectively. For a fixed temperature, the radiation of natural bodies is reduced by the emissivity, thus $T_B < T$. In the use of radiometer for the surface temperature, the detected surface layer is less than 1 cm for liquid water but more for dry ice or snow. Interpretation of data can be difficult in the melting period, since surface layer is isothermal, and the emissivity of a pixel is influenced by both ice and meltwater. A drawback of satellite radiometry is the poor spatial resolution, kilometres at best.

Return backscatter of radars from snow and ice is sensitive mostly to the surface roughness, relative permittivity, liquid water content, porosity, and grain structure. These properties may change fast that makes interpretation of radar imagery difficult, and usually it requires extra information. SAR technique provides a high-resolution satellite method by creating a synthetic antenna using the satellite movement (e.g., [26]).

With wavelengths of centimetres, the penetration depth of SARs may be several metres in dry ice but much less in wet ice. Radar can recognize different kinds of ice but because of the large difference in the relative permittivity between the liquid and solid phases of water, radar is highly sensitive to the wetness of the surface layer. The sensitivity to the surface roughness concerns particularly the length scale of the radar wavelength. Very good results have been obtained for displacements of surface features, such as sea ice drift, from consecutive images. A specific technique is SAR interferometry, which provides information of the surface topography and small lateral displacements.

Airborne long wavelength (0.3–3 m) radars, the so-called ground penetrating radars or georadars, can penetrate to ice and consequently be useful for collecting information about the vertical structure of glaciers (e.g., [39]). These radars are often mounted on a surface-vehicle such as a sledge.

Study Problems

1. What is the freezing point of freshwater at the surface of Lake Vostok, located 4 km beneath the surface of the Central Antarctic ice sheet?
2. How long can conical icicle grow before its own weight breaks it?
3. Consider the steady state of heat conduction equation (Eq. 2.13): $\frac{d}{dz}\left(k\frac{dT}{dz}\right) + \frac{dQ_s}{dz} = 0$ where $Q_s = Q_{s0}e^{\lambda z}$, $z = 0$ at the surface. Determine the depth of the temperature maximum.
4. If the irradiance spectrum is even, what is the relation between the quantum irradiance and irradiance power integrated over the PAR band?
5. What must be distance of geostationary satellites from Earth's surface?

References

1. Petrenko VF, Whitworth RW (1999) Physics of ice. Oxford University Press
2. Pounder E (1965) Physics of ice. Pergamon Press
3. Murray BJ, Knopf DA, Bertram AK (2005) The formation of cubic ice under conditions relevant to earth's atmosphere. Nature 434:202–205
4. Pauling L (1935) The structure and entropy of ice and other crystals with some randomness in atomic arrangement. J Am Chem Soc 57:2680–2684
5. Michel B (1978) Ice mechanics. Laval University Press
6. Eicken H, Gradinger R, Salganek M, Shirasawa K, Perovich D, Leppäranta M (eds) (2009) Field techniques in sea ice research. University of Alaska Press
7. Langway CC (1958) Ice fabrics and the universal stage. Techn. Rep. 62. U.S. Army Snow, Ice, and Permafrost Research Establishment, Hanover NH
8. Leppäranta M, Kosloff P (2000) The thickness and structure of Lake Pääjärvi ice. Geophysica 36(1–2): 233–248
9. Moreira S, Schultze M, Rahn K, Boehrer B (2016) A practical approach to lake water density from electrical conductivity and temperature. Hydrol Earth Syst Sci 20:2975–2986
10. Nakawo M (1980) Density of columnar-grained ice made in a laboratory. National Research Council of Canada. Tech. rep. 168
11. Hunter SC (1976) Mechanics of continuous media. Ellis Horwood
12. Mase GE (1970) Continuum mechanics. McGraw-Hill, Schaum's outline series
13. Ashton G (ed) (1986) River and lake ice engineering. Water Resources Publications
14. Mellor M (1986) Mechanical behavior of sea ice. In: Untersteiner N (ed) The geophysics of sea ice. Plenum Press, pp 165–281
15. Cuffey KM, Paterson WSB (2010) The physics of glaciers, 4th edn. Academic Press
16. Leppäranta M (2011) The drift of sea ice, 2nd edn. Springer-Praxis
17. Palmer A, Croasdale K (2012) Arctic offshore engineering. World Scientific Publishing
18. Kundu PK, Cohen IM (2004) Fluid mechanics, 3rd edn. Elsevier
19. Mellor M, Cole DM (1982) Deformation and failure of ice under constant stress or constant strain-rate. Cold Reg Sci Technol 5(3):201–219
20. Schulson EM, Duval P (2009) Creep and fracture of ice. Cambridge University Press
21. Glen JW (1958) The flow law of ice. Int Assoc Sci Hydrol 47:171–183
22. Davis RO, Selvadurai APS (2002) Plasticity and geomechanics. Cambridge University Press
23. Coon MD (1980) A review of AIDJEX modeling. In: Pritchard RS (ed) Proceedings of ICSI/AIDJEX symposium on sea ice processes and models. University of Washington Press, pp 12–23
24. Sree DKK, Law AW-K, Shen HH (2017) An experimental study on the interactions between surface waves and floating viscoelastic covers. Wave Motion 70:195–208
25. Leppäranta M, Lindgren E, Arvola L (2016) Heat balance of supraglacial lakes in the western Dronning Maud land. Ann Glaciol 57(72):CJO2016
26. Rees WG (2006) Remote sensing of snow and ice. CRC Press
27. Dera J (1990) Marine physics. Elsevier
28. Kirk JTO (2011) Light and photosynthesis in aquatic ecosystems, 3rd edn. Cambridge University Press

29. Warren SG (2019) Optical properties of ice and snow. Phil. Trans. R. Soc A377:20180161
30. Lu P, Leppäranta M, Cheng B, Li Z, Istomina L, Heygster G (2018) The color of melt ponds on Arctic Sea ice. Cryosphere 12:1331–1345
31. Shaw A, Declerq NF (2015) Determination of ice thickness on lakes and canals using audio sound. Acta Acust united Acust 101(3):484–493
32. Tedesco M (ed) (2014) Remote sensing of the cryosphere. Wiley
33. Voss KJ, Chapin A, Monti M, Zhang H (2000) Instrument to measure the bidirectional reflectance distribution function of surfaces. Appl Opt 39(33):6197–6206
34. Riggs GA, Hall DK, Román MO (2015) VIIRS snow cover algorithm. Report. NASA Goddard Space Flight Center
35. Del Grande NK, Clark GA, Durbin PF, Fields DJ, Hernandez JE, Sherwood RJ (1991) Buried object remote detection technology for law enforcement. SPIE Surveillance Technol 1479:335–351
36. Steffen K (1985) Surface temperature and sea ice of an Arctic polynya: north water in winter. In: Zürcher Geographische Schriften, vol 19. ETH Geographisches Institute
37. Leppäranta M, Lewis JE (2007) Observations of ice surface temperature in the Baltic Sea. Int J Remote Sens 28(17):3963–3977
38. Carsey FD (Ed.) (1992) *microwave remote sensing of sea ice*. Geophysical monograph series 68, Am Geophys Union
39. Woodward J, Burke MJ (2007) Applications of ground-penetrating radar to glacial and frozen materials. J Environ Eng Geophys 12:69–85

3 Atmosphere in Cold Regions

*Halla [night frost, in Finnish]. Occurrence of night frost in summer is a threat to harvest. Night frost caused famine years in North Europe especially in the Little Ice Age. Painting from 1895 by Hugo Simberg (1873–1917). (*Finnish National Gallery, *Photograph Jenni Suominen)*

The Earth's atmosphere and its interactions with cryosphere is presented. The text begins with introducing the basic atmospheric variables and their distributions with elevation followed by solar and terrestrial radiation. Then the surface mass balance is introduced, particularly the

M. Leppäranta, P. Uotila, *Geophysics of the Cryosphere*, Springer Textbooks in Earth Sciences, Geography and Environment, https://doi.org/10.1007/978-3-031-93071-3_3

formation of snow crystals and snowfall. The last section presents the basic atmospheric boundary layer physics and surface heat balance. After carefully reading this chapter, a student can comprehend the importance of the atmosphere for cryosphere and solve the snow an ice physics problems in the chapters to follow.

3.1 Atmosphere

3.1.1 Structure of Atmosphere

The Earth's atmosphere contains nitrogen (78%), oxygen (21%), argon (0.9%), other gases such as carbon dioxide (0.04%), aerosol particles, and water in all its phases (e.g., [1]). The vertical structure shows layers with the troposphere closest to the surface (to 6–20 km), and then the stratosphere (up to 50 km). Weather processes take place in the troposphere which is at highest at the equator and lowest at the poles (e.g., [1–3]). Stratosphere contains ozone (O_3) which protects the Earth from dangerous ultraviolet radiation (UV-A and B), and occasionally polar stratospheric clouds form there as the only weather phenomenon.

The principal atmospheric variables are air pressure, temperature, humidity, wind, cloudiness, and precipitation. Air is taken as an ideal gas. The equation of state is

$$p_a = \rho_a R_d T_a \tag{3.1}$$

where p_a is the air pressure, ρ_a is the air density, R_d= 287 J K^{-1} kg^{-1} is the gas constant of dry air,[1] and T_a is the air temperature. Air pressure decreases with increasing height according to the hydrostatic pressure law

$$\frac{dp_a}{dz} = -\rho_a g \tag{3.2}$$

where g is the acceleration due to gravity, and z is the vertical coordinate positive up. The sea surface pressure is between 900 and 1100 mbar (or hPa, hectopascal), and its standard value is 1013.25 mbar. At this pressure and $T_a = T_f$ (freezing point) or $T_a = T_f - 20$ K, $\rho_a = 1.29$ kg m^{-3} or $\rho_a = 1.39$ kg m^{-3}, respectively.

Air temperature is a widely used atmospheric characteristic in geosciences and ecology. It is quite stable, easy to measure, and long-term records are available for many observation stations around the world. The annual and daily cycles of air temperature follow the solar radiation, and in general, their amplitudes are larger in dry than in wet areas. In practice, when air temperature is above the freezing point, ice may form only in very specific conditions such as surface hoar deposition during night-time radiative cooling.

Example 3.1 Time integrals of air temperature, the so-called *degree-day sums*, are key characteristics in climatology. They are defined as the sums of the daily mean temperatures $T_{a,\,k}$ with respect to a given reference. For example, $\sum_k \max(T_f - T_{a,\,k}, 0\ °C)$ is the sum of freezing-degree-days.

Water vapour pressure (e) is the partial pressure caused by water vapour molecules in air. Just above a water surface, water molecules are continuously entering or leaving the surface. When air is saturated with water vapour, the net flux of water molecules is zero, and the pressure is called the *saturation water vapour pressure*, e_s. It increases with temperature, $e_s = e_s(T)$, following the Clausius–Clapeyron equation

$$\frac{de_s}{dT} = \frac{L_e(T)e_s}{R_v T^2} \tag{3.3}$$

where L_e is the latent of evaporation (at liquid water surface) or sublimation (at ice surface), and R_v is the gas constant of water vapour. Due to the temperature dependence of latent heat, an analytic solution does not exist but there are good approximations available (Fig. 3.1; for an explicit formula, see Annex A2.3). At temperatures below the freezing point, saturation pressure is given both for liquid water and ice surfaces, and $e_{s,\,ice} < e_{s,\,liquid}$. In cold environment, the saturation

[1] Gas constant = universal gas constant/molecular mass of gas.

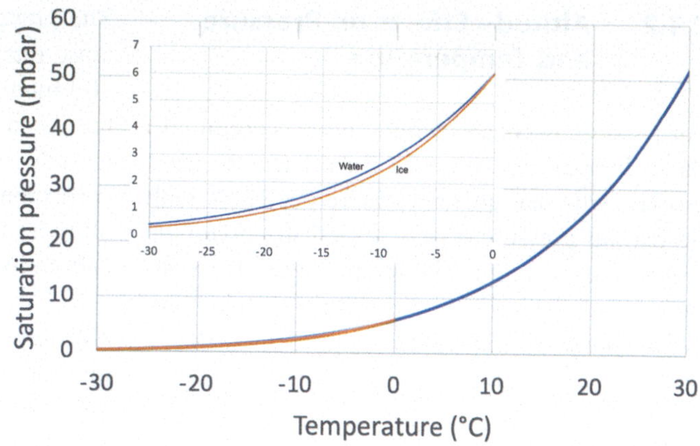

Fig. 3.1 Saturation water vapour pressure as a function of temperature at water surface and at ice surface from −30 °C to 30 °C. The close-up is from −30 °C to 0 °C showing the difference between ice and water surface

water vapour pressure is small that limits precipitation and sublimation/deposition; $e_s(T_f) = 6.11$ mbar.

Atmospheric water vapour content can be given by water vapour pressure or *specific humidity q*, the mass fraction of the water vapour.[2] They are related by $q = 0.622e/p_a$. *Relative humidity R* is the fraction of the absolute humidity from the saturation level,

$$R = \frac{q}{q_s} = \frac{e}{e_s} \tag{3.4}$$

where q_s is the saturation specific humidity. For example, at the standard atmospheric pressure, $q_s(T_f) = 3.75$ g/kg. Annual variations in relative humidity are usually small, while absolute humidity largely follows the air temperature. The *dew point/frost point* T_D is defined as the temperature where condensation/sublimation takes place for the given humidity and temperature:

$$e_s(T_D) = Re_s(T_a) \tag{3.5}$$

Always $T_D \leq T_a$, and surface hoar formation occurs when $T_D < T_f$.

Wind is driven by the atmospheric pressure field and Earth's rotation (the apparent Coriolis force) and, in the planetary boundary layer, by the friction at the Earth's surface [4]. The balance between the pressure gradient and the Coriolis force gives the *geostrophic wind*, which blows along the isobars with high pressure on the right (left) in the Northern (Southern) Hemisphere with respect to the blow direction. The *surface wind* is slower, blowing by 10–30° from isobars towards low pressure. Westerly winds dominate in 30–60° latitudes, while polar easterlies exist in high latitudes. Clouds are classified for their type and sky coverage or *cloudiness*, given in oktas. Cloudiness is a critical factor in the radiation balance, and clouds hide the Earth's surface in optical and infrared satellite remote sensing. Accumulated *precipitation* is measured by precipitation gauges in millimetres of liquid water over time. The world average annual precipitation is about 1000 mm, smaller over continents than over the ocean, and smaller in cold climate regions.

In normal weather data, air pressure is expressed as its value at the sea surface level using an altitude correction (see next subsection). Air temperature and humidity are provided at the standard 2 m altitude, while the standard altitude of surface wind is 10 m. Wind speed is reported as a 10-minute average, while gusts represent 3-s averages. Weather data may include cloud classification and the type of precipitation. There are extensive atmospheric data stores such as the global ERA5 of the European Centre of Medium-Range Weather Forecasts (ECMWF) with analysed hourly data since the year 1940 in 0.25° × 0.25° lateral resolution.

[2] Absolute humidity is the mass of water vapor per volume of air, usually expressed in g/m^3.

3.1.2 Altitude Effects on Pressure and Temperature

Air pressure decreases with altitude according to the hydrostatic equation (Eq. 3.2). Then, the ideal gas law tells that air temperature decreases with decreasing pressure, that is called the *adiabatic temperature change*. The temperature of dry air brought adiabatically to the surface is the *potential temperature*, expressed as

$$\theta = T\left(\frac{p}{p_0}\right)^{R_d/c_p} \tag{3.6}$$

where p_0 is the surface pressure, c_p is the specific heat of dry air at constant pressure, and $R_d/c_p \approx 0.286$. This model is useful in the troposphere but in the stratosphere, due to the absorption of ultraviolet solar radiation, air temperature increases with altitude.

The stability of stratification tells whether a vertically displaced parcel returns (stable), escapes (unstable), or stays in the new position (neutral). In a neutrally stratified atmosphere, the potential temperature is constant, while the actual temperature decreases by the *adiabatic lapse rate*

$$\Gamma \equiv -\frac{\partial T}{\partial z} = \frac{g}{c_p} \tag{3.7}$$

In dry air, $\Gamma = 9.8$ K km^{-1}, for moist air, it is less because condensation and deposition release heat and partly compensate for the influence of the pressure release. In the troposphere, the standard adiabatic lapse rate is $\Gamma = 6.5$ K km^{-1} (e.g., [4, 5]). The stratification is stable (unstable) when the actual lapse rate is smaller (larger) than the adiabatic lapse rate.

The vertical distribution of air pressure is obtained by the hydrostatic equation and ideal gas law. Assuming a constant lapse rate, the result is:

$$p(z) = p_0\left(1 - \frac{\Gamma z}{T_0}\right)^{\frac{g}{R_d \Gamma}} \tag{3.8}$$

The power in Eq. (3.8) is 5.26 for the standard lapse rate, and the scale of pressure reduction is 100 mbar km^{-1} in the troposphere. Air pressure has been used as an approximate measure for altitude, since 1 mbar pressure difference corresponds to 10 m in altitude. For the absolute altitude, the pressure at sea level is needed for calibration.

Example 3.2 Assume standard atmospheric pressure, $T_a = T_f = 273.15$ K at the sea level, and neutral stratification. The density of air is 1.293 kg m^{-3} at sea level. At 5 km altitude, the air pressure is 520.4 mbar, $T_a = 240.7$ K, and the density of air is 0.782 kg m^{-3}.

The decrease of temperature with altitude shows up in the cryosphere boundaries. High mountains provide good environmental conditions for the existence of snow and ice, and therefore glaciers exist even in tropical regions. The low local air pressure has an impact on the solubility of gases in water, but the freezing point and temperature of maximum density of surface water are affected by the pressure drop by less than 0.01 °C, even in high mountains. Thin atmosphere above mountains has influence on the radiation balance, and high in the atmosphere, cold temperature supports snow crystallization, precondition of snowfall.

3.1.3 Cold Climate Zones

The world's climate conditions are classified according to the Köppen–Geiger system into five main classes depending on the monthly mean temperature and precipitation (Table 3.1; Fig. 3.2). In natural environment, apart from certain exceptional cases, the existence of snow and ice necessitates the surface air temperature to be below the freezing point, that is possible in snow climate, polar climate, and in parts of arid and warm temperate climates. Arid climates are further subclassified, and two coldest ones may contain ice. Perennial ice exists only in polar climate

Table 3.1 Köppen–Geiger system of climate classification; T is monthly mean temperature, $\tilde{T}$ is annual mean temperature, and P is precipitation [6]

A	Equatorial climates	$T_{min} > 18\ ° C$
B	Arid climates Cold arid climate subclasses – Cool summer and cold winter – Extremely continental	$P < 10P_{th}$ [a] $\tilde{T} < 18\,° C$ $T_{min} > -38\ ° C$ $T_{min} \leq -38\ ° C$
C	Warm temperate climate	$-3\ ° C < T_{min} < 18\ ° C$
D	Snow climates	$T_{min} < -3\ ° C$
E	Polar climates	$T_{max} < 10\ ° C$

[a] $P_{th} = 2\tilde{T}$ ($P_{winter} \geq \frac{2}{3}P$), $2\tilde{T} + 28$ $\left(P_{summer} \geq \frac{2}{3}P\right)$, or $2\tilde{T} + 14$ (otherwise); units mm, °C

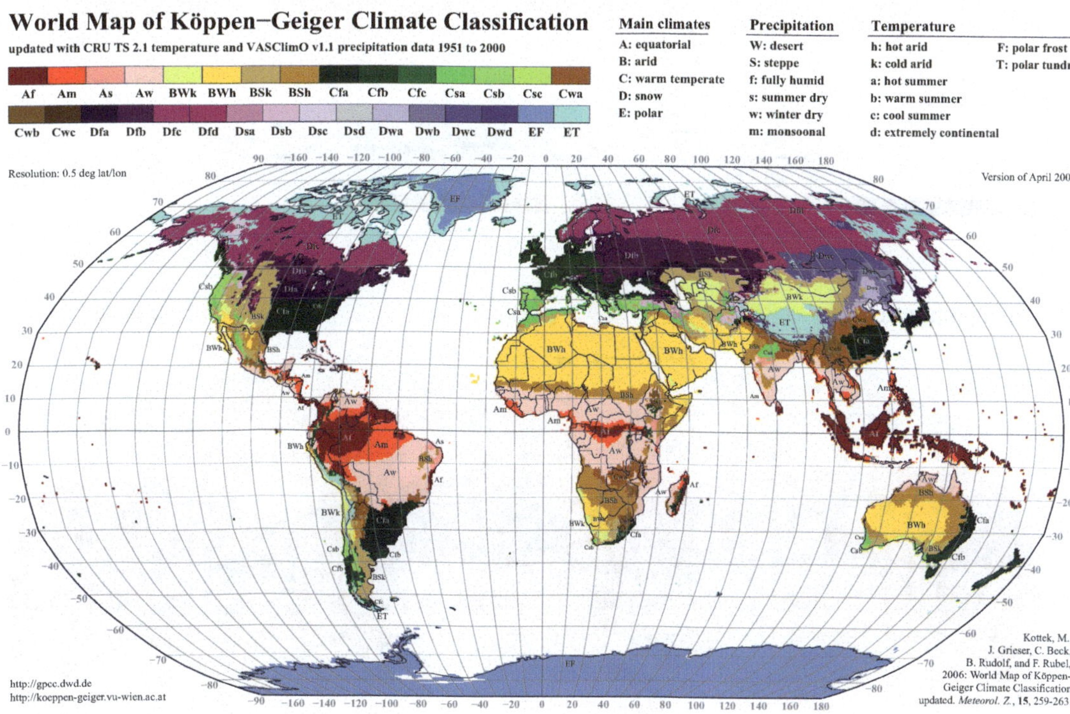

Fig. 3.2 Köppen–Geiger classification of climates [6]. Printed by permission

in ice sheets and glaciers, permafrost, and multi-year sea ice or lake ice.

Figure 3.3 shows the Earth's mean air temperature in January and July. In January it is less than the freezing point over most of Eurasia and North America north from 40 to 45°N latitude at sea level. The boundary is more north in western Europe due to the warm North Atlantic current and more south in the Central Asian mountains. In the Southern Hemisphere, apart from Antarctica, snow and ice areas near the sea level altitude are found in small Southern Ocean islands, at about 40°S in New Zealand South Island and Tasmania, and in the South American tip reaching 55°S. In lower southern latitudes, cold regions exist in the Andes.

Large-scale atmospheric circulation characteristics are described by indexes based on the atmospheric pressure distribution. North Atlantic Oscillation (NAO) index is based on the height difference in the 500-mbar atmospheric pressure surface between Ponta Delgada, Azores

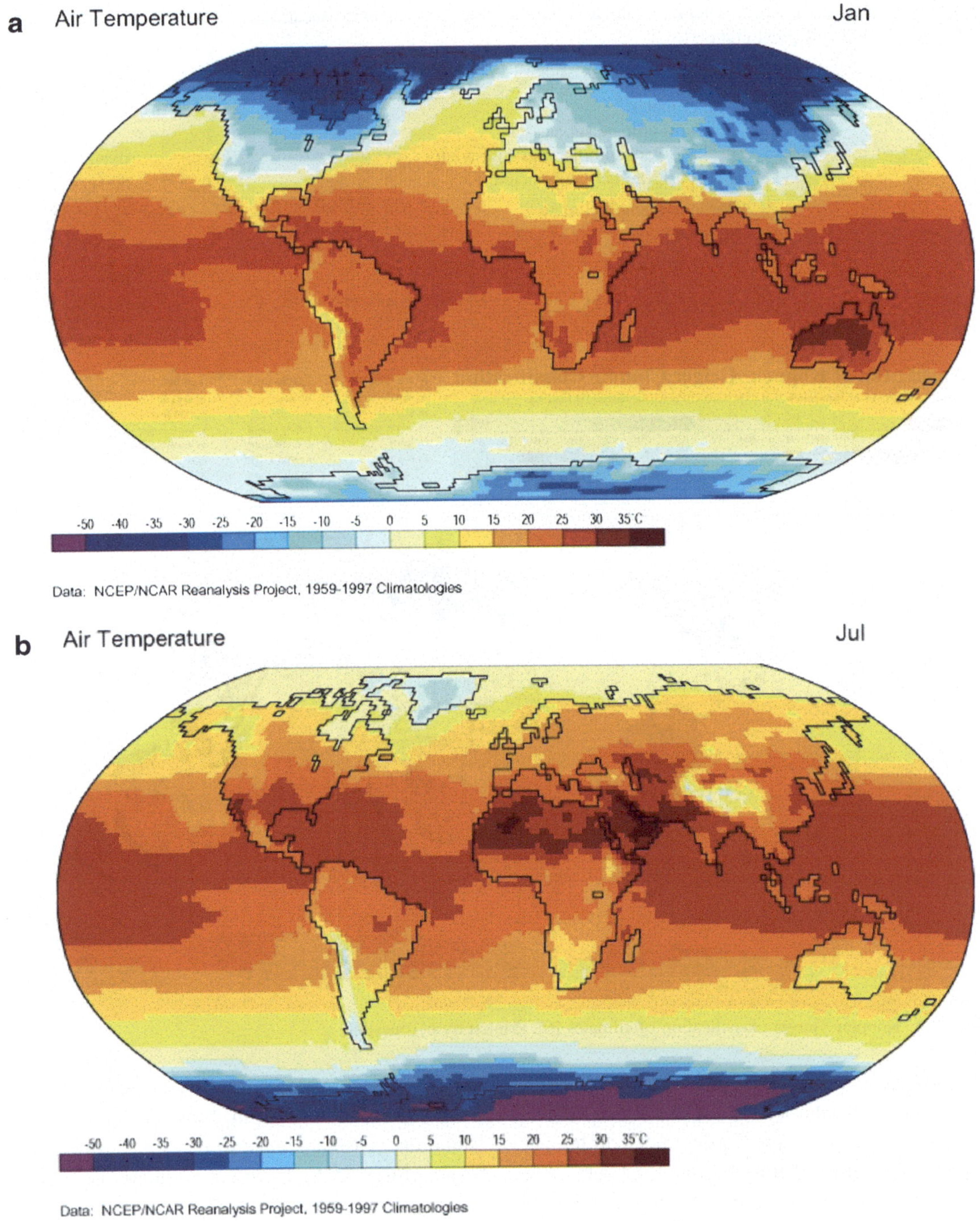

Fig. 3.3 Mean surface air temperature (°C) in **a** January and **b** July 1959–1997. Global climate animations, Department of Geography, University of Oregon based on NCEP/NCAR Reanalysis Project 1959–1997 Climatologies data. © Department of Geography, University of Oregon

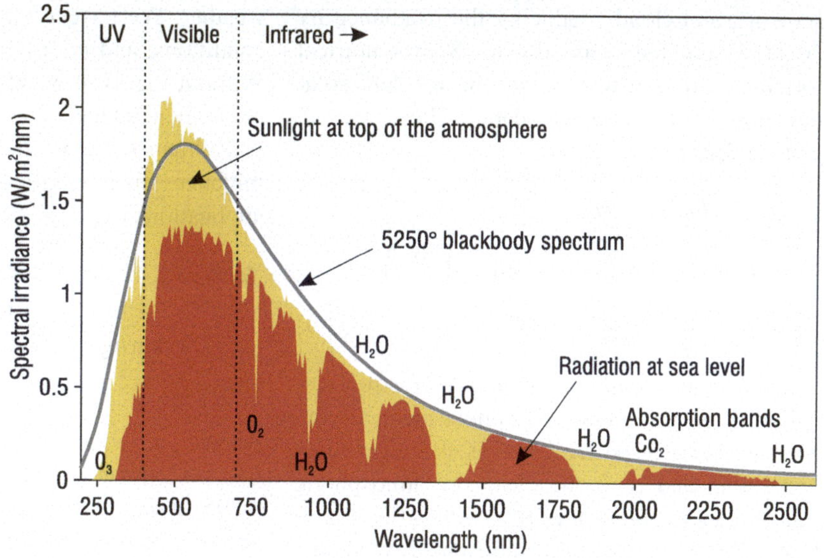

Fig. 3.4 Solar radiation spectrum at the top of the atmosphere and at the sea level. Atmospheric absorption bands due to ozone (O_3), oxygen (O_2), water vapour and carbon dioxide (CO_2) are illustrated in the solar radiation band

and Stykkishólmur, Iceland, defined as the normalized anomaly of this difference. NAO is a key index in Europe telling of the intensity of the westerlies. When the index is positive, westerly winds are stronger than normal and winters in Europe are warm, and vice versa. In the land regions around the Pacific Ocean, El Niño–Southern Oscillation (ENSO) is connected to anomalously warm and cold winters. NAO cannot be predicted, but ENSO has a significant intraseasonal memory. For example, it has been utilized in seasonal forecasting of ice winters for the Great Lakes of North America [7].

3.2 Solar and Terrestrial Radiation

3.2.1 Solar Radiation

The Sun radiates almost as a black body (Eq. 2.19) at the temperature of its outer surface, 5900 K, and the solar spectrum (Fig. 3.4) has nearly all its energy in the 300–3000-nm wavelength window [8]. The spectrum peaks at 480 nm and shows Fraunhofer lines in the optical band due to absorption by elements in the solar atmosphere. The solar heat flux is obtained by integration of the spectrum over the wavelengths. The *solar constant* $Q_{sc} = 1367$ W m^{-2} is the annual average power of solar radiation incident on a plane perpendicular to the solar rays on the top of the atmosphere (e.g., [8]). According to satellite measurements, the solar constant has varied within 0.1% in recent decades. Due to the ellipticity of Earth's orbit, the actual extra-terrestrial radiation has annual cycle from 1324 W m^{-2} in summer solstice to 1418 W m^{-2} in winter solstice.

At the Earth's surface, the optical solar spectrum is quite even, i.e., sunlight is white. The ultraviolet bands UV-A and UV-B are absorbed by ozone in the stratosphere. Oxygen absorption peak shows up at 760 nm, and water vapour has near-infrared absorption peaks at 950 nm and 1100 nm. Beyond 1200 nm, water vapour and carbon dioxide have several absorption bands. Incident direct solar radiation on a horizontal plane on Earth's surface depends on the solar altitude angle ($h_{\odot}$) which in turn depends on time of day, season, and location, obtained from astronomical formulae (see Annex A2.2).

Incident solar irradiance on a horizontal plane on Earth's surface[3] (Q_s) consists of direct solar beam and diffuse radiation from the whole sky. Attenuation of radiation in the atmosphere is due to absorption and scattering by gas molecules,

[3] This is also called global radiation.

aerosols, and clouds, reducing the irradiance by 20–80% on the way down. Semi-empirical formulae are used to estimate the incident solar radiation from weather data. They are in general form:

$$Q_s(0^+; h_\odot, e, N, r_E) = F(N, h_\odot)\cdot \tau_{tr}(h_\odot, e)\cdot \sin h_\odot \left(\frac{AU}{r_E}\right)^2 Q_{sc} \quad (3.9)$$

where N is the cloudiness, r_E and AU (astronomical unit) are the actual and annual mean Earth–Sun distance, respectively, F is the cloudiness correction, and τ_{tr} is the clear sky atmospheric transmittance. The factors $\tau_{tr} \approx 0.8 - 0.9$ and F, $0.3 \precsim F \leq 1$, represent the atmosphere, while the last, astronomical factor depends only on time and location.

Example 3.3 At spring and autumn equinoxes, declination is zero, daytime is 12 h, and solar altitude is $\sin h_\odot = \cos\phi \cos\tau$, where ϕ is the latitude and τ is the hour angle (see Annex A2.2 for the formulas). At noon, $\tau = 0$,and the solar zenith angle ($Z = 90° - h_\odot$) equals the latitude, $Z = \phi$. The daily mean of $\sin h_\odot$ is

$$\langle \sin h_\odot\rangle = \frac{1}{2}\left(\frac{1}{2}\int_{-\pi/2}^{\pi/2} \cos\phi \cos\tau \, d\tau\right) = \frac{1}{2}\cos\phi$$

Thus, at $\phi = \pm 60°$, radiation is reduced from the solar constant by the astronomical factor of 0.25. Since the atmospheric reduction scales to 0.5, the daily mean radiation is about 10% of the solar constant, as seen in observations in Finland [9].

Ångström–Prescott equation provides a practical way to estimate the daily mean radiation [10]:

$$Q_s = F\cdot\tau_{tr}\cdot\tilde{Q}_{sE} = (a_s + b_s\nu_s)\cdot\tilde{Q}_{sE} \quad (3.10)$$

where a_s~$0.10 - 0.30$ and b_s~$0.30 - 0.60$ are empirical parameters, ν_s is the fraction of sunshine in the daytime hours, and $\tilde{Q}_{sE}$ is the mean extra-terrestrial radiation of the day. The parameters a_s and b_s depend on the location and weather. We have $F\cdot\tau_{tr} \approx a_s + b_s$ in clear sky conditions, and $F\cdot\tau_{tr} \approx a_s$ in fully cloudy days. Without the local weather data, Allen et al. [10] recommended the values $a_s = 0.25$ and $b_s = 0.50$.

Zillman's [11] formula for the clear sky transmittance has been found good in many applications (e.g., [5, 12]):

$$\tau_{tr}(h_\odot, e) = \frac{\sin h_\odot}{1.085\sin h_\odot + e\cdot(2.7 + \sin h_\odot)\cdot 10^{-3} + 0.01} \quad (3.11)$$

where e is given in millibars. At $e = 6.11$ mbar $= e_s(T_f)$, we have $\tau_{tr} = 0.804 - 0.895$ at $h_\odot = 10°–90°$. At very small angles, accuracy is low; the limit $h_\odot \to 0$ gives $\tau_{tr} \to 0$ that is not exactly true since diffuse radiation is received at sunset.

The cloud correction depends on the amount and quality of clouds, and its variability and uncertainty are larger there than in the clear sky transmittance. Usually only the total cloudiness is available. Lumb [13] examined the hourly incident radiation at different cloudiness levels and cloud types. Several formulae have been designed for the daily solar flux, a widely used form is [14]:

$$F\left(N, \widehat{h}_\odot\right) = 1 - c_N N + c_\odot \widehat{h}_\odot \quad (3.12)$$

where $\widehat{h}_\odot$ is the solar altitude at noon in degrees, $c_N = 0.62$ is the cloudiness coefficient, and $c_\odot = 0.0019$ is the solar angle coefficient. It is seen that $F \geq 0.38$, and the formula must be cut to satisfy $F \leq 1$ (see [5, 14]).

The net solar radiation (Q_{sn}) absorbed by the Earth's surface is a fraction of the incident solar radiation,

$$Q_{sn} = (1 - \alpha)Q_s \quad (3.13)$$

where α is the albedo (see Eq. 2.24). Part $Q_r = \alpha Q_s$ is reflected and scattered back to the atmosphere. Albedo depends on the properties of the medium and the directional distribution of the incident radiation (see Sect. 2.4.2). It is about 0.9 for dry snow, 0.5 for dry ice, and 0.07 for liquid water (e.g., [15]). Snow and ice albedos have large variability: both decrease with increasing liquid water content in the scattering layer, snow albedo decreases with increasing grain size, and ice albedo increases with increasing gas content.

Net solar radiation is further divided into two parts: absorption of infrared radiation in the 5–10 cm near-surface layer (Q_{s0}) and penetration of sunlight and near-UV radiation deeper into the medium (Q_{s+}):

$$Q_{s0} = \gamma Q_{sn}, Q_{s+} = (1-\gamma)Q_{sn}, \gamma \sim 0.5 \quad (3.14)$$

Q_{s0} goes into the surface boundary condition, and Q_{s+} acts as a source term in the heat conduction equation (see Eq. 2.13).

3.2.2 Terrestrial Radiation

Earth surface radiates nearly as a black body following the Planck's law (Eq. 2.19), reduced by the grey body emissivity ε_0, $0 < \varepsilon_0 < 1$. At the Earth's surface temperature, energy emission takes place in the thermal infrared band (3–100 μm). Since solar and terrestrial radiation spectra have practically no overlap, they have been also called *shortwave* and *longwave radiation*, respectively. Integrating the grey body spectrum over the wavelengths gives the energy flux (see Eq. 2.20)

$$Q_{L0} = \varepsilon_0 \sigma T_0^4 \quad (3.15)$$

where σ is the Stefan–Boltzmann constant, and T_0 is the surface temperature. The emissivity of snow, ice, and water varies only little between 0.96 and 0.98. The radiation loss from the surface is 180–340 W m^{-2} when the surface temperature varies between 240 K and 280 K.

Estimation of the thermal radiation of the atmosphere is complicated because the radiation comes from different altitudes with different temperatures and emissivities. In the common semi-empirical formula, the complexity is disguised in an effective emissivity, for which semi-empirical formulae are used based on humidity and cloudiness, $\varepsilon_a = \varepsilon_a(e, N)$, and in the air temperature at 2-m level acting as the reference temperature:

$$Q_{La} = \varepsilon_a \sigma T_a^4 \quad (3.16)$$

Atmospheric radiation is available also in atmospheric analysis data and in model output.

A common approach is to take the emissivity as the product $\varepsilon_a = \varepsilon_{a1}(e)\varepsilon_{a2}(N)$ of clear sky emissivity and cloud cover correction as

$$\varepsilon_{a1} = A + B\sqrt{e} \quad (3.17a)$$

$$\varepsilon_{a2} = 1 + CN^2 \quad (3.17b)$$

The clear sky case (Eq. 3.17a) is from Brunt [16], and typical parameter values are $A = 0.68$ and $B = 0.036$ mbar$^{-1/2}$; A represents the dry atmosphere emissivity, while the square root reflects the vertical distribution of water vapour. The cloudiness correction has been used in northern Europe with $C = 0.18$ [12, 17]. As in the case of solar radiation, cloudiness is the most uncertain factor. Based on Eq. (3.17b), the minimum emissivity is $A = 0.68$. At $T = T_f$, the clear sky emissivity with the saturation water vapour pressure is 0.77, and with a full cloud cover added, the emissivity is 0.91. The emissivity is $\varepsilon_a \sim 0.7 - 0.9$, and for the air temperature between 240 K and 280 K, $Q_{La} = 130 - 310$ W m^{-2}.

3.3 Surface Mass Balance

3.3.1 Snow in the Atmosphere

Snow crystals form in clouds. Adiabatic cooling of rising air increases relative humidity, and when the water vapour reaches the saturation level, condensation into liquid droplets or, in low temperature, deposition into snow crystals takes place. Liquid droplets may supercool in very cold air until they contact with nuclei and freeze. Atmospheric circulation dynamics, orography, and surface heating may trigger the necessary upward motion of air. Snow and ice crystals fall onto Earth's surface as solid precipitation when the temperature of the atmospheric boundary layer is low for the solid phase survival. Fast falling hail may succeed in falling through warm atmosphere in summer. Snowfall measurements with precipitation gauges are challenging in turbulent conditions. Attempts have been made to improve the data quality by gauge design (Fig. 3.5), accounting for the aerodynamics, but in general data tend to be biased downward.

Water vapour is invisible, but droplets and crystals form the visible clouds in the sky. Since the saturation vapour pressure is lower for deposition than condensation (Fig. 3.1), snow crystals are likely to form when the temperature is below the freezing point. This takes place as heterogeneous nucleation on atmospheric aerosols such as dust and pollen. The primary snow crystals grow by secondary nucleation and retain symmetrical shape as its growing arms experience the same atmospheric conditions.

Snow crystals are unique forms of ice. Their size and shape show a huge variability, as illustrated by the phrase 'no two snowflakes are alike' [18]. The morphology of snow crystals depends on the air temperature and degree of supersaturation, as illustrated by the Nakaya's diagram (Fig. 3.6, see also Portrait 3.1). The most beautiful snowflakes form when the air temperature is between −15 °C and − 10 °C and the air humidity is weakly supersaturated. More than 120 snowflake types have been identified and classified, from those with hexagonal symmetry, such as six-armed beauties and symmetric plates, into more irregular columnar and needle-type forms, such as graupels, pellets, and hail [20, 21]. When snow crystals deposit on ground, they form a layer with distinct thermal and mechanical properties as discussed further in Chap. 4 on seasonal snow cover.

Fig. 3.5 Tretyakov-model precipitation gauge, metal shade slices are around for aerodynamic control of snow accumulation. (*Photograph* Matti Leppäranta)

In conditions of cooling of water droplets, such as rising air in thunderstorms, the droplets freeze to form typically spherical ice crystals. Ice pellets form from refreezing of melted snow crystals, while snow pellets (Fig. 3.7) are soft white balls with supercooled water nucleating on snow crystals. The diameter of pellets is less than 5 mm. Hail forms in thunderstorms with strong rising motion of air, and the ice crystals, also called hail stones, are larger than ice pellets, even several centimetres in diameter. Hail stones falling at high speed can reach the ground in summer, occasionally causing damage when their size is large.

3.3.2 Freezing of Liquid Precipitation

Since the saturation water vapour pressure is small in cold climate, also precipitation is usually low. Heavy precipitation occurs in mid-latitude mountains, such as the Scandinavian Mountain Chain in North Europe and Rocky Mountains on the west coast of North America, with annual precipitation up to 3000 mm, whereas over the Antarctic Plateau precipitation is as in tropical deserts, less than 100 mm in a year. Precipitation data may include the type, whether it is rain, showers, drizzle, hail, snow, wet snow, or freezing rain. In climate modelling, usually the volume and phase are needed. A practical approach has been to take precipitation solid when the air temperature is below a specified level, e.g., 1 °C.

Liquid precipitation results in frozen land surfaces in particular conditions. A minor case is

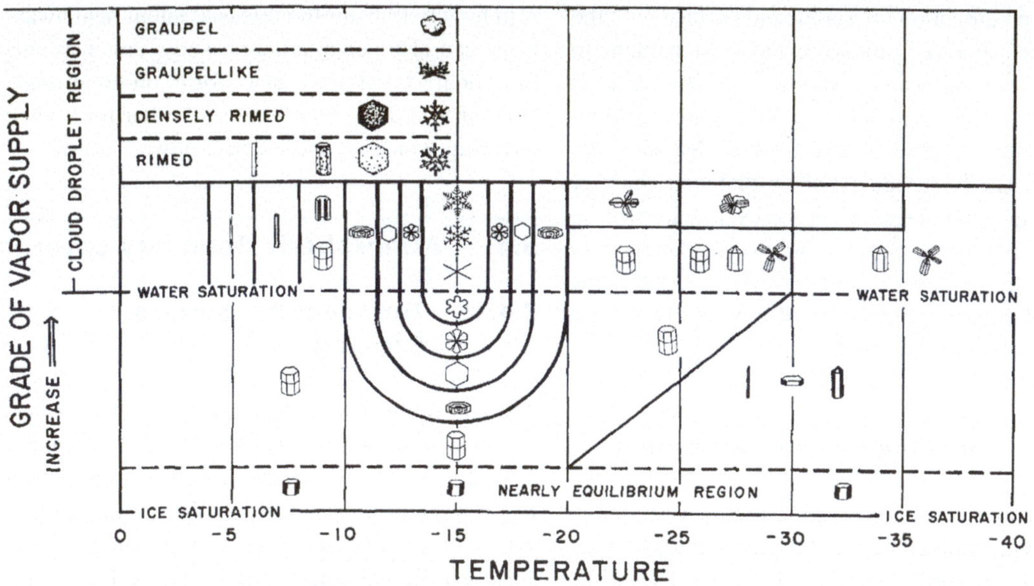

Fig. 3.6 Shapes of snow crystals in clouds and their dependence on temperature and vapour supply (Nakaya's diagram) [19]

Portrait 3.1 Ukichiro Nakaya (1900–1962) was a pioneering snow scientist. He majored in experimental physics at Tokyo Imperial University, graduating in 1925. In 1930 he began with snow research in Hokkaido University and received his Doctor of Science degree from Kyoto Imperial University. He founded the Institute of Low Temperature Science in Hokkaido University in 1935, and a year later he was the first to create artificial snowflakes in his laboratory. He continued snow research and contributed to atmospheric ice accretion and glaciology. In 1954, his summarizing book *Snow crystals: Natural and artificial* was published. His well-known quote is *Snowflakes are letters sent from heaven. Wikimedia Commons.* (*Photograph* unknown author (1946))

Fig. 3.7 Snow pellets on ground in southern Finland, May 2020. (*Photograph* Matti Leppäranta)

liquid precipitation on a cold land surface with the sub-surface layer providing the cold content to freeze the rainwater. Another, drastic case is *freezing rain*, when supercooled liquid precipitation falls on ground and freezes. Massive ice layers can be produced then that can damage buildings and trees. Such cases are known to occur in Eastern Siberia and North America in connection of Arctic inflows. For example, the city of Montreal, Québec suffered badly from a heavy ice storm in 1998.

3.3.3 Sublimation and Deposition

When the surface temperature is below the freezing point, sublimation, and deposition transfer ice between the atmosphere and Earth's surface. The direction of the transfer depends on the difference in the water vapour pressure between the surface and air,

$$\Delta e = e(T_a) - e_s(T_0) \tag{3.18}$$

If $\Delta e > 0$, deposition of water vapour occurs on the surface, and if $\Delta e < 0$ and $e(T_a) < e_s(T_a)$, sublimation takes place. The thickness of the sublimated/deposited layer, Δh, is limited by the available heat flux:

$$\Delta h = \frac{Q_e t}{\rho L_E} \tag{3.19}$$

where Q_e is the latent heat flux, t is the time, and ρ is the density of the surface layer. Thus, sublimation by 1 mm ice per day at the freezing point temperature needs a heat flux of 30.0 W m^{-2}; $Q_e > 100$ W m^{-2} would be an exceptionally high value in cold environment. However, in dry and cold climate regions, sublimation can become a significant term in the mass balance, order of 1–5 mm day^{-1} [22, 23].

Deposition of water vapour on land and vegetation surface, *frost*, is a common consequence in cold clear nights when longwave radiation loss results in a low surface temperature. Beautiful crystal forms are called frost flowers. A limitation is in how to transfer the released latent heat away. Frost can also accumulate on snow cover as surface hoar. Frost may also form on man-made structures and the outcome can be harmful when covering windows and electric power cables.

3.4 Atmospheric Boundary Layer

3.4.1 Turbulent Air–Surface Exchange

The atmospheric boundary layer is in direct interaction with the Earth's surface. Its scale height is 1 km, and atmospheric near-surface measurements are usually made in the lowest 10 m layer. The interaction is dictated by the surface characteristics, and boundary layer temperature, humidity, wind, and stratification are examined using the turbulent boundary theory with different orders of turbulence closure (e.g., [24, 25]).

Turbulent transfer is larger by several orders of magnitude compared to molecular diffusion. Turbulent fluxes are maintained by eddies generated by wind shear and buoyancy. In a turbulent flow, the time history of any quantity is set as the sum of the average and fluctuation parts, known as the Reynolds decomposition. For example, velocity is $\boldsymbol{u} = (u, v, w) = (\overline{u}, \overline{v}, \overline{w}) + (u', v', w')$, and potential temperature is $\theta = \overline{\theta} + \theta'$, where overbar marks for the time average and apostrophe marks for fluctuation. The fluctuations are white noise with zero time average.

According to the boundary layer theory, the vertical turbulent flux of any quantity is proportional to its covariance with the vertical velocity, the so-called eddy covariance [25]. The fluxes of momentum, sensible heat and latent heat are defined positive downward, by, respectively,

$$\tau = -\rho_a \overline{u'w'} \tag{3.20a}$$

$$Q_H = -\rho_a c_p \overline{\theta' w'} \tag{3.20b}$$

$$Q_E = -\rho_a L_E \overline{q'w'} \quad (3.20c)$$

where x-axis is aligned with the wind, and q' is for fluctuations in the specific humidity. These fluxes are assumed constant in the surface boundary layer. They can be obtained directly with eddy covariance measurement techniques, or they can be estimated from boundary layer mean data with different orders of turbulence closure. The first-order closure includes the gradient flux method and the bulk formula.

3.4.2 Gradient Flux Method

The method is based on the Fick's diffusion law which assumes that the flux is proportional to the gradient of the quantity in question. This is analogous to molecular transfer and introduces the concept of *eddy diffusivity K*, dim K = length^2/time (see [24]). For the potential temperature,

$$\overline{\theta' w'} \sim K \frac{d\overline{\theta}}{dz} \quad (3.21)$$

Eddy diffusivity is a property of the flow, not a property of the fluid, and it varies over a wide range. By dimensional analysis reasoning, it depends on the turbulent velocity and length scales u_* and ℓ, respectively:

$$K \sim u_* \ell \quad (3.22)$$

where u_* is called the *friction velocity*, and in the first-order theory, $u_* \propto |\overline{u} - \overline{u}_0|$, where $\overline{u}$ is the mean wind speed and $\overline{u}_0$ is the mean speed of the surface (usually in ice and snow applications, $\overline{u} - \overline{u}_0 \approx \overline{u}$ since $|\overline{u}| \gg |\overline{u}_0|$). According to Prandtl's *mixing length hypothesis,*[4] the size of turbulent eddies in the boundary layer scales with the distance from the boundary, $\ell \propto z$. The mixing length hypothesis is based on shear production of turbulence with no buoyancy effects, i.e., strictly speaking neutral stratification, but the model is applicable in near-neutral conditions.

[4] Prandtl's mixing length hypothesis assumes that turbulent transfer processes are described by a characteristic length scale over which the turbulent eddies mix fluid properties.

The assumption of constant fluxes in the surface layer, e.g., $u_* z \frac{d\overline{u}}{dz} = const.$, leads to the logarithmic profiles [25, 26]:

$$\overline{u}_z - \overline{u}_0 = \frac{u_*}{\kappa} \log\left(\frac{z}{z_0}\right) \quad (3.23a)$$

$$\overline{\theta}_z - \overline{\theta}_0 = \frac{\theta_*}{\kappa} \log\left(\frac{z}{z_\theta}\right) \quad (3.23b)$$

$$\overline{q}_z - \overline{q}_0 = \frac{q_*}{\kappa} \log\left(\frac{z}{z_q}\right) \quad (3.23c)$$

where $\kappa = 0.4$ is von Kármán constant, θ_* and q_* are the turbulent temperature and humidity scales, and z_0, z_θ, and z_q are the roughness lengths for momentum, temperature, and humidity, respectively. These roughness lengths follow from the surface boundary conditions ($\overline{u}_z - \overline{u}_0 = 0$ at $z = z_0$, etc.) and do not exactly represent the geometric roughness. In the 10-m surface layer, the potential temperature nearly equals the actual temperature (see Eq. 3.6), the difference is less than 0.01 °C. On snow, ice, and water surface, humidity is at the saturation level, $q_0 = q_s(\theta_0)$.

3.4.3 Bulk Formulas

In a non-stratified flow over a smooth surface, turbulent momentum transfer should depend on the density and velocity of the fluid only. By dimensional analysis, we have $\tau \propto \rho u^2$. With similar reasoning for heat and moisture, we have the bulk formulas

$$\tau = \rho_a C_{Dz} (\overline{u}_z - \overline{u}_0)^2 \quad (3.24a)$$

$$Q_H = \rho_a c_p C_{Hz} (\overline{\theta}_z - \overline{\theta}_0) |\overline{u}_z - \overline{u}_0| \quad (3.24b)$$

$$Q_E = \rho_a L_E C_{Ez} (\overline{q}_z - \overline{q}_0) |\overline{u}_z - \overline{u}_0| \quad (3.24c)$$

where C_D, C_H, and C_E are dimensionless bulk exchange coefficients. In near-neutral stratification, these coefficients depend on the reference height of atmospheric data, and surface roughness. For z = 10 m, $C_D = C_H = C_E = 1.2 \cdot 10^{-3}$ is a representative value over snow and ice (e.g., [27]). The absolute values of the sensible and latent heat fluxes are usually less than 50 W m^{-2}.

Example 3.4 Assume near-neutral stratification and take the $u = 5\ \mathrm{m\ s}^{-1}$, $T_a = -5\ ^\circ C$, $T_0 = -2\ ^\circ C$, and $R = 80\%$ at 10 m altitude. Momentum transfer is 0.039 N m^{-2}, sensible heat flux is -23.5 W m^{-2}, and latent heat flux is -26.6 W m^{-2} ($e = 0.8 \cdot e_s(T_a) = 3.21$ mbar, $e_0 = e_s(T_0) = 5.17$ mbar $> e$).

The gradient flux and bulk formulas have similar structure since

$$K\frac{d\overline{\theta}}{dz} \approx u_* z \frac{\overline{\theta}_z - \overline{\theta}_0}{z} \propto (\overline{u}_z - \overline{u}_0)(\overline{\theta}_z - \overline{\theta}_0) \tag{3.25}$$

With $\rho_a u_*^2 = \rho_a C_{Dz} \overline{u}_z^2$, etc., it is seen that the bulk transfer coefficients and surface roughness lengths are dependent through

$$\sqrt{C_{Dz}} \log\left(\frac{z}{z_0}\right) = \sqrt{C_{Hz}} \log\left(\frac{z}{z_\theta}\right) = \sqrt{C_{Ez}} \log\left(\frac{z}{z_q}\right) = \kappa \tag{3.26}$$

Thus, the neutral bulk coefficients for snow and ice surfaces C_D, C_H, $C_E = 1.2{\cdot}10^{-3}$ ($z = 10$ m) correspond to the roughness lengths of $z_0 = z_\theta = z_q = 0.1$ mm. Ten-folding of the roughness changes the bulk coefficients by about 50%.

Example 3.5 The standard observation levels are $z_2 = 2$ m for the temperature and humidity and 10 m for the wind. In near-neutral conditions, the temperature is at the 10 m level

$$\overline{\theta}_{10} - \overline{\theta}_0 = \frac{\log(10\ \mathrm{m} - z_\theta)}{\log(2\ \mathrm{m} - z_\theta)} (\overline{\theta}_2 - \overline{\theta}_0)$$

The log-ratio is 1.16 and 1.21 for the roughness lengths of 0.1 mm and 1 mm, respectively.

3.4.4 Stratified Flow

Turbulent fluxes are largely influenced by the stratification in the atmospheric boundary layer that is treated in the Monin–Obukhov theory (see, e.g., [25]). Stratification brings the buoyancy length scale to the problem, and therefore the bulk transfer coefficients become dependent on the stability of stratification. If the stratification is stable (unstable), turbulent fluxes are reduced (strengthened). For example, in polar winter cold air descends into valleys and stable stratification develops, while unstable situation arises with cold air flow over a warm lake during autumn cooling. Compared to near-neutral case, in highly stable stratification the bulk transfer coefficients can be lower by more than 50%, and in highly unstable stratification, they are higher by up to 25%.

The stability of stratification is expressed by the Richardson number, which equals the ratio between buoyant and shear production of turbulence (e.g., [24, 25]). The flux Richardson number Ri_f and the gradient Richardson number Ri_G are defined as

$$Ri_f = \frac{g}{\theta_v} \cdot \frac{\overline{\theta' w'}}{u_*^2\, \partial\overline{u}/\partial z} \approx \frac{g}{\theta_v} \cdot \frac{\partial\overline{\theta}/\partial z}{(\partial\overline{u}/\partial z)^2} = Ri_G \tag{3.27}$$

where $\theta_v = (1 + 0.608q)\overline{\theta}$ is the virtual temperature. The eddy diffusivities of momentum and temperature are assumed equal in deriving Ri_G. Stratification is stable (unstable) for $Ri_f > 0$ ($Ri_f < 0$). In the stable case, turbulence is resisted by negative buoyancy and damped out when $Ri_f > 0.2$. In practice, the bulk Richardson number Ri_B is often used, where the gradients in Ri_G are approximated by finite differences:

$$Ri_B = \frac{gz}{\theta_v} \cdot \frac{\overline{\theta}_z - \overline{\theta}_0}{(\overline{u}_z - \overline{u}_0)^2} \approx Ri_G \tag{3.28}$$

In general form, the bulk exchange coefficients depend on the altitude, surface roughness, and stability: $C_{Dz}(z_0, Ri)$, $C_{Hz}(z_\theta, Ri)$, $C_{Ez}(z_q, Ri)$. These coefficients are interconnected via the Richardson number, and therefore the turbulent fluxes (Eqs. 3.24) need to be solved simultaneously. The solution cannot be given in explicit form, but iterative procedures are used (see [26–28]). The full solution is quite technical, not presented here. Instead, approximate explicit formulations are available providing an insight

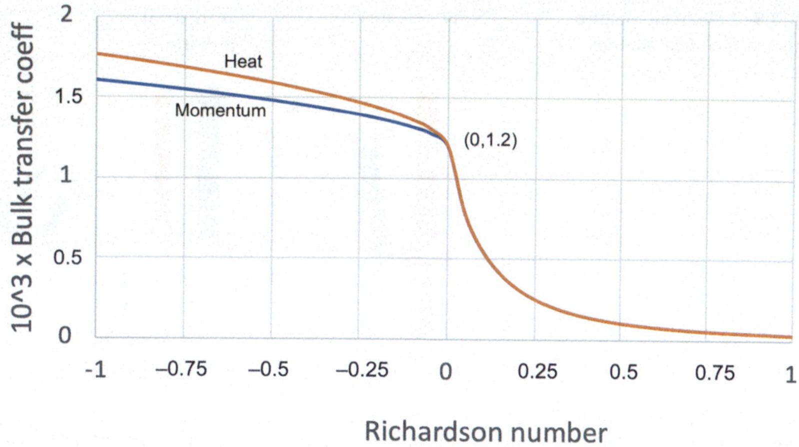

Fig. 3.8 The turbulent bulk transfer coefficients of momentum (C_D) and sensible (C_H) and latent (C_E) heat vs. stability of stratification according to the explicit equations of Louis [29]. The neutral ($Ri = 0$) value is $C_D = C_H = C_E = 1.2 \cdot 10^{-3}$, and in the unstable side ($Ri < 0$), $C_D < C_H = C_E$

to the role of stratification. Louis [29] parameterization ($z \gg z_0$ and $z_0 = z_\theta = z_q$ assumed) reads:

$$C_D = C_H = C_E = \left[\frac{1}{\kappa}\log\left(\frac{z}{z_0}\right)\right]^{-2} f(Ri_B) \tag{3.29a}$$

$$Ri_B > 0 : f(Ri_B) = (1 + 4.7Ri_B)^{-2} \tag{3.29b}$$

$$Ri_B < 0 : f(Ri_B) = 1 - \frac{9.4Ri_B}{1 + B^* C_D(z_0, 0)\sqrt{\frac{z}{z_0}|Ri_B|}} \tag{3.29c}$$

where f is the stability function, and $B^* = 70$ for momentum and 50 for heat and moisture fluxes (Fig. 3.8). For example, take $z = 10$ m and $z_0 = 0.1$ mm. The neutral transfer coefficient is $C_D = 1.2 \cdot 10^{-3}$. Then for $Ri_B = -0.2$, $C_D = 1.38 \cdot 10^{-3}$, and $C_H = C_E = 1.45 \cdot 10^{-3}$, and for $Ri_B = 0.2$, $C_D = C_H = C_E = 0.32 \cdot 10^{-3}$.

3.5 Surface Heat Balance

3.5.1 Heat Budget

The heat content of a snow or ice surface layer per unit surface area consists of latent and sensible heat. It is convenient to have liquid water at the freezing point temperature T_f as the reference level. Then the heat content relative to the reference is obtained by integration:

$$\Delta\mathscr{H} = -\int_{-h}^{0} (1-\nu)\rho_i\left[L_f + c_i\left(T_f - T\right)\right]dz \tag{3.30}$$

where h is the layer depth, and ν is is the porosity of ice. Eventual water is here at the freezing point and does not add to the heat content. The time evolution of the heat content is obtained directly from the change of porosity and ice temperature using Eq. (3.30) or from the driving fluxes (Fig. 3.9):

$$\frac{d}{dt}\Delta\mathscr{H} = Q_0 + Q_{s+}\left[1 - \exp\left(-\int_{-h}^{0}\kappa dz\right)\right] + Q_d + Q_A \tag{3.31}$$

where Q_0 is the surface heat balance, Q_d is the heat flux from the bottom of the layer, and Q_A is lateral heat transfer. Positive net flux is seen as temperature increase and melting. The direct and flux methods can be used to control each other (e.g., [30]).

3.5.2 Surface Heat Balance

The surface heat balance consists of infrared solar radiation Q_{s0}, terrestrial radiation $Q_{La} - Q_{L0}$, turbulent sensible Q_H and latent Q_E heat fluxes, and heat transfer via precipitation, Q_P (Fig. 3.9). Snow, ice, and liquid water are transparent to sunlight, but infrared band is absorbed in the near-surface layer and added in the surface heat

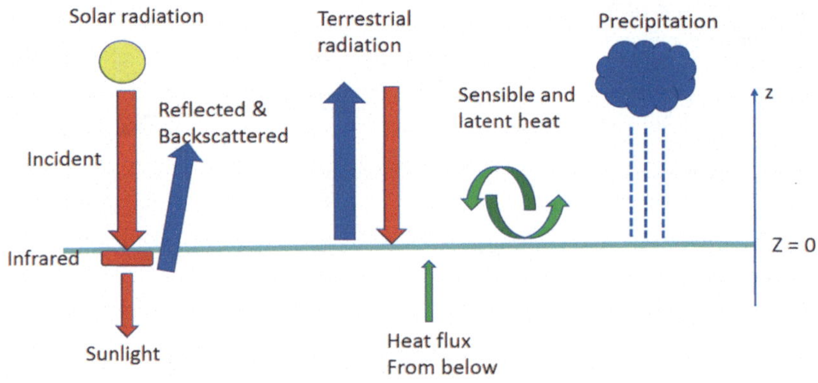

Fig. 3.9 Schematic picture of the surface heat budget of a snow or ice layer

balance (see Eq. 3.14). The surface heat balance is written (e.g., [2, 5])

$$Q_0 = Q_{s0} + Q_{La} - Q_{L0} + Q_H + Q_E + Q_P \tag{3.32}$$

The net surface heat flux is balanced by conduction from/to the ice below and eventual melting/freezing:

$$\rho L_f \frac{\partial h}{\partial t} + k \frac{\partial T}{\partial z}\bigg|_{z=0^-} + Q_0 = 0; T_0 \le T_f \tag{3.33}$$

In the case of melting, another question is that what happens to meltwater: storage, evaporation, or advection away.

The radiation balance $Q_R = Q_{sn} + Q_{La} - Q_{L0}$ is the governing factor in Earth's heat budget. In the polar regions, solar radiation has a strong annual cycle between polar day and polar night, from zero to 500 W m^{-2} in daily values, and albedo is a major question in snow and ice environment. The net terrestrial radiation, $Q_{La} - Q_{L0}$ is mostly negative and scales as −50 W m^{-2} in cold regions. Positive values may occur in extreme cases, e.g., when there is warm air advection over melting ice [30].

Heat transfer from precipitation contains sensible heat due to the temperature difference between precipitation and the surface, and latent heat due to eventual phase change. The latter, i.e., snowfall on open water or rain on snow or ice can be an important transient heat transfer mechanism. For example, for 10 mm day^{-1} rain falling on 5 °C colder water, the sensible heat flux is $\rho_w c_w \Delta T = 2.4$ W m^{-2}, but for the same rate of snowfall falling on water and melting, the heat flux is $\rho_w L_f P = -38.6$ W m^{-2} which is comparable to the turbulent heat fluxes. Snow cover is permeable to liquid water, and therefore raindrops can transport heat below the surface. In the boreal zone, the monthly average precipitation is around 50 mm, and therefore precipitation remains a small factor in the monthly climatology.

Example 3.6 A planetary radiation balance can be written as.

$$\frac{1}{4}(1-\alpha)Q_s = \varepsilon_0 \sigma T_0^4$$

when the contribution of atmosphere is neglected. The factor ¼ is because solar radiation falls on a disc of area πR^2 while the area of a spherical planet is $4\pi R^2$, where R is the planet radius. The temperature solution is $T_0 = \sqrt[4]{\frac{(1-\alpha)Q_s}{4\varepsilon_0\sigma}}$. For Earth, the planetary albedo is 0.3, $\varepsilon_0 = 0.97$, and Q_s is the solar constant, and we have $T_0 = 257$ K. The real average is 288 K, so the greenhouse effect of the present atmosphere is about 30 K.

3.5.3 Linearized Heat Balance

In climate research and mathematical modelling, a linearized form of the surface heat flux is advantageous since first-order solutions can be obtained for a general overview [31]. The linearized surface heat flux is expressed as

$$Q_0 = K_0 + K_1(T_a - T_0) \quad (3.34)$$

where the coefficients K_0 and K_1 do not depend explicitly on T_0. They are obtained from the radiative and turbulent fluxes presented in Sects. 3.2 and 3.4 based on the standard weather variables: air temperature, relative humidity R, wind speed u_a, and cloudiness N.

Solar radiation is absorbed in K_0 either the net value $Q_{sn} = (1 - \alpha)Q_s$ or the surface layer absorption Q_{s0} only when sunlight Q_{s+} is taken as a source term inside the medium, depending on the application. Terrestrial radiation is split into two terms using the binomial formula for the temperature:

$$T_0^4 = [T_a + (T_0 - T_a)]^4 \approx T_a^4 + 4T_a^3(T_0 - T_a) \quad (3.35)$$

and the net terrestrial radiation is then written in linear form as

$$\Delta Q_L \approx -\varepsilon_0\sigma(1 - \varepsilon_a)T_a^4 + 4\varepsilon_0\sigma T_a^3(T_0 - T_a) \quad (3.36)$$

The first term in the right-hand side is the zeroth-order terrestrial radiation balance, where $T_0 = T_a$, and adds on K_0; and the second term is the first-order correction for $T_0 \neq T_a$, and the coefficient $4\varepsilon_0\sigma T_a^3$ is included in K_1. The truncated formula is accurate to about 1% when $|T_a - T_0| < 10\,°$ C and $T_a \sim 0\,°$ C.

The sensible heat flux is proportional to the temperature difference $T_a - T_0$, and thus the proportionality coefficient goes to K_1. The latent heat exchange is split using the Clausius-Clapeyron equation for the saturation water vapour pressure e_s and assuming that humidity is at the saturation level at the surface, $e_o = e_s(T_0)$. We have

$$Q_e = -\rho_a L_E C_E \frac{0.622 e_s(T_a)}{p_a} \times \left[(1 - R) - \frac{L_E}{R_v T_a^2}(T_a - T_0)\right] U_a \quad (3.37)$$

Combining all together, the terms in the coefficients K_0 and K_1 are shown in Table 3.2. These expressions look complicated, but the resulting climatological values are rather smooth. With seasonal variation coming largely from the solar radiation and the latent heat flux. It is seen that $K_0 \gtreqless 0$ is dominated by the radiation balance, and $K_1 \geq 4\varepsilon_0\sigma T_a^3 > 0$. We can call K_0 as the zeroth-order heat balance which holds exact when $T_0 = T_a$.

Study Problems

1. Show that solar altitude at noon is $\theta_0 = 90\,° - \phi + \delta$, where ϕ is latitude and δ is declination.
2. Examine using the radiation balance whether the 'snowball Earth' is a possible equilibrium.
3. (a) Determine the heat loss in evaporation of 1 mm water or sublimation of 1 mm ice in 1 day at 0 °C. (b) The relative humidity is 50%, wind speed is 10 m s^{-1}, and surface temperature and air temperature are equal. Determine sublimation at the temperatures of $0\,°\,C$ and $-20\,°\,C$.
4. Net heat loss from an ice layer is 50 W m^{-2}. Assuming that the heat loss is distributed evenly in a 5 m thick surface layer, estimate how much the temperature decreases in this layer in 2 days.
5. Consider an open water spot in a frozen lake. Determine the surface heat balance when air temperature is a) 10 °C and b) –10 °C, and relative humidity is 70%, wind speed is 5 m s^{-1}, cloudiness 3/8, and net solar radiation is 100 W m^{-2}.

Table 3.2 The terms of the coefficients K_0 and K_1 in the linearized surface heat balance $Q_0 = K_0 + K_1(T_a - T_0)$

	Solar radiation	Terrestrial radiation	Sensible heat	Latent heat
K_0	$Q_{s0} + Q_{s+}$ Or Q_{s0}	$-\varepsilon_0\sigma(1 - \varepsilon_a)T_a^4$	–	$\rho_a L_E C_E \frac{0.622 e_s(T_a)}{p_a}(1 - R)U_a$
K_1	–	$4\varepsilon_0\sigma T_a^3$	$\rho_a c_a C_H U_a$	$\rho_a L_E C_E \frac{0.622 e_s(T_a)}{p_a} \frac{L_E}{R_v T_a^2} U_a$

References

1. Wallace JM, Hobbs PV (2006) Atmospheric science, an introductory survey, 2nd edn. Academic Press
2. Peixoto JP, Oort AH (1992) Physics of climate. American Institute of Physics
3. Vallis G (2019) Essentials of atmospheric and oceanic dynamics. Cambridge University Press
4. Holton JR, Hakim GJ (2013) An introduction to dynamic meteorology, 5th edn. Academic Press
5. Curry JA, Webster PJ (1999) Thermodynamics of atmospheres and oceans. Academic Press
6. Kottek M, Grieser J, Beck C, Rudolf B, Rubel F (2006) World map of the Köppen-Geiger climate classification updated. Meteorol Z 15(3):259–263
7. Wang J, Bai X, Hu H, Clites A, Holton M, Lofgren B (2012) Temporal and spatial variability of Great Lakes ice cover, 1973–2010. J Clim 25:1318–1329
8. Iqbal M (1983) An introduction to solar radiation. Academic Press
9. FMI (1982) Solar radiation measurements 1971–1980. In: Meteorological Yearbook of Finland, Part 4:1 Measurements of solar radiation 1971–1980. Finnish Meteorological Institute
10. Allen RG, Pereira LS, Raes D, Smith M (1988) Crop evapotranspiration—guidelines for computing crop water requirements. FAO Irrigation and drainage paper 56. FAO
11. Zillman JW (1972) A study of some aspects of the radiation and heat budgets of the Southern Hemisphere oceans (562 p.). In: Meteorological studies, vol 26. Bureau of Meteorology, Department of Interior, Canberra
12. Leppäranta M, Myrberg K (2009) Physical oceanography of the Baltic Sea. Springer-Praxis
13. Lumb FE (1964) The influence of cloud on hourly amounts of total solar radiation at the sea surface. Q J R Meteorol Soc 90(383):43–56
14. Reed RK (1977) On estimating insolation over the ocean. J Phys Oceanogr 7:482–485
15. Perovich DK (2016) Sea ice and sunlight. In: Thomas DN (ed) Sea ice, 3rd edn. Wiley, pp 110–137
16. Brunt D (1932) Notes on the radiation in the atmosphere. Q J R Meteorol Soc 58(247):389–420
17. Omstedt A (1990) A coupled one-dimensional sea ice-ocean model applied to a semi-enclosed basin. Tellus 42A:568–582
18. Nakaya U, Hanajima M, Muguruna J (1958) Physical investigations on the growth of snow crystals. J Facul Sci Ser. II V(3):87–133
19. Magano C, Lee CW (1966) Meteorological classification of natural snow crystals. J Fac Sci Hokkaido Univ Ser. VII 2(4):321–335
20. Fierz C, Armstrong RL, Durand Y, Etchevers P, Greene E, McClung DM, Nishimura K, Satyawali PK, Sokratov SA (2009) The international classification for seasonal snow on the ground. In: IHP-VII technical documents in hydrology. IACS contribution (1), vol 83. UNESCO-IHP, Paris
21. LaChapelle ER (1992) A field guide to snow crystals. International Glaciological Society
22. Huang W-F, Li Z, Han H, Niu F, Lin Z, Leppäranta M (2012) Structural analysis of thermokarst lake ice in Beiluhe Basin, Qinghai–Tibet Plateau. Cold Reg Sci Technol 72:33–42
23. Leppäranta M, Lindgren E, Arvola L (2016) Heat balance of supraglacial lakes in the western Dronning Maud land. Ann Glaciol 57(72):39–46
24. Kundu PK, Cohen IM (2004) Fluid mechanics, 3rd edn. Elsevier
25. Tennekes H, Lumley JE (1972) A first course in turbulence. MIT Press
26. Kagan BA (1995) Ocean-atmosphere interaction and climate modelling. Cambridge University Press
27. Andreas EL (1998) The atmospheric boundary layer over polar marine surfaces. In: Leppäranta M (ed) The physics of ice-covered seas, vol 2. Helsinki University Press, pp 715–773. https://helda.helsinki.fi/handle/10138/39286
28. Launiainen J, Vihma T (1990) Derivation of turbulent surface fluxes—an iterative flux-profile method allowing arbitrary observing heights. Environ Softw 5(3):113–124
29. Louis J-F (1979) A parametric model of vertical eddy fluxes in the atmosphere. Bound-Layer Meteorol 17: 187–202
30. Leppäranta M, Lindgren E, Wen L, Kirillin G (2019) Ice cover decay and heat balance in Lake Kilpisjärvi in Arctic tundra. J Limnol 78(2):163–175
31. Leppäranta M (2023) Freezing of lakes and the evolution of their ice cover, 2nd edn. Springer

4 Seasonal Snow Cover

In North Europe, moose hunting was based on skis from the stone age until modern times. A hard crust forms in the surface layer of snowpack in late winter that bears a skier but not a moose. (Painting by Väinö Hämäläinen (1876–1940) illustrates a Finnish folklore story from the epic *Kalevala*. © *Helsingin Suomalainen Yhteiskoulu)*

This chapter presents the geophysics of the seasonal snow. Snow crystals form in the atmosphere, fall due to gravity, and accumulate onto the Earth's surface producing the seasonal snow cover. This white, soft layer is a specific winter environment, and it protects the ground from cold and light. Snow–atmosphere interaction and snow metamorphic processes bring major changes to snowpack. The lifetime of seasonal snow is, by definition, less than 1 year, and at the final melting, seasonal snow returns in liquid phase to the global water cycle. Surviving over the summer,

M. Leppäranta, P. Uotila, *Geophysics of the Cryosphere*, Springer Textbooks in Earth Sciences, Geography and Environment, https://doi.org/10.1007/978-3-031-93071-3_4

Portrait 4.1 Gerald Seligman (1886–1973) was a snow scientist and founder of the International Glaciological Society. He was born in London, went to the Natural Sciences Tripos at Trinity Hall, Cambridge, where he graduated in 1908. As a snow personality, he changed from a skier and Alpinist to scientist. In 1931, he resigned from business to devote himself to scientific work that was consolidated when he spent a year at the Scott Polar Research Institute in Cambridge, writing his influential book *Snow structure and ski fields* which was published in 1936. After World War II, he transformed the Association for the Study of Snow and Ice into the International Glaciological Society, and in 1951–1954, he was President of the Commission of Snow and Ice. In 1963, the Glaciological Society decided to institute an award for the occasional honouring of individuals of particular eminence in the field of glaciology, to call this award the Seligman Crystal. (*Photograph* Archives of the International Glaciological Society. Printed by permission)

seasonal snow becomes firn[1] and forms a layer of perennial land ice treated in Chap. 6 (Portrait 4.1).

4.1 Physical Properties of Seasonal Snow

4.1.1 Snow Cover

Snow is a specific form of ice. We can characterize snow cover by its white colour, low density, softness, and permeability (e.g., [1, 2]). The white appearance is due to the strong and spectrally even scattering and weak absorption of sunlight in the snowpack. Low-density results from falling snow crystals joining together into a loose, soft net with open structure that has finite strength and is permeable for gases. For example, diffusion of water vapour in snowpack is one key factor in snow metamorphosis, and transport of oxygen and carbon dioxide are important to snow ecosystems.

Seasonal snow cover is a laterally continuous or patchy medium on the Earth's surface (Fig. 4.1). Snow depth is equal to the distance from the ground to the snow surface, with the scale from centimetres near the climatological snow margin to more than 1 m in large areas in the boreal and tundra climate. The deepest snow is several metres in mid-latitude mountains with heavy precipitation, such as the Rocky Mountains on the west coast of North America [3] and the Japanese Alps in the island Honshu. In high polar climate, snow accumulation is not large due to low precipitation, despite of the long winter. In the Antarctic Plateau, precipitation is less than 100 mm per year, as in the Sahara Desert.

4.1.2 Fundamental Properties of Snow

The fundamental properties of a snowpack are the depth, density, grain (crystal) size and shape, and liquid water content (Table 4.1). Characteristic to these properties is their large variability in time and space. Snow properties data are collected from snow pits (Fig. 4.2) based on the international standards presented in Fierz et al. [4]. Liquid water content can be significant only at the freezing point temperature T_f that makes 'warm snow' ($T = T_f$) in many respects a different medium from cold snow ($T < T_f$).

The density of new snow is near 100 kg m^{-3} which means that 90% of the volume is air and 10% is ice crystals. The vertically averaged density increases in snow metamorphic processes in winter's progress but low-density layers within a snowpack may develop. The density of snow changes by wind packing, pressure by the

[1] Firn is old snow; the word originates from German 'firn' which means last year's or previous.

Fig. 4.1 Snow cover over Europe and northwestern Asia in winter 2001–2002 based on MODIS images. Product of NASA Goddard Space Flight Center, Scientific Visualization studio

Table 4.1 Scales of the fundamental physical properties of seasonal snow [2, 4–6]

Property	Scale	Comments
Depth	0.1–1 m	Several metres in mid-latitude mountains
Density	100–400 kg m^{-3}	New snow 100 kg m^{-3}; boreal forest snow 200–250 kg m^{-3}; wind-packed tundra snow 300–400 kg m^{-3}
Grain size	1–10 mm	Change due to metamorphic processes
Grain shape	Rich variety of types	
Liquid water content, Volume fraction	Wet snow 0^{+}–5% Slush >5%	Significant at the melting point temperature Water holding capacity of snow 5%

overlaying snow, sintering,[2] sublimation and water vapour deposition, and snow melt–freeze processes. There are density evolution models forced by the weather history [6–8]. The density is usually measured by weighing a snow sample of a fixed volume. The empirical threshold density between seasonal snow and firn is 550 kg m^{-3} which corresponds to the density of randomly packed spheres[3] of ice.

Snow grains are first as the falling snow crystals. They become rounded in strong wind, or, settled on the ground, their fine branches break, and they freeze together in metamorphic processes. The grain size is typically within 1–10 mm, and the grains are classified based on their size and shape [4]. The grain structure has

[2] Bonding of snow grains together at temperatures below the melting point has been termed *sintering*, in analogy with the phenomenon known in powder metallurgy.

[3] Random packing of equal-size ice spheres gives a volume fraction of 0.64 or a density of 587 kg m^{-3}.

Fig. 4.2 Working in a snow pit. The stratigraphy is recorded from the wall, and the profile is quantified with in situ data and samples. (*Photograph* Matti Leppäranta)

been studied by photographic and 3D tomographic methods (e.g., [9]). A common statistic to characterize the grain size of a sample is the maximum diameter.

The liquid water content, given by the volume fraction ν_w, is a delicate and critical quantity in warm snow. In cold snow $\nu_w \approx 0$, but when the temperature has reached the melting point, additional heat produces melt water which is first adhered to snow crystals. The moisture can be felt when taking a snow sample in hand, e.g., to make a snowball. The water holding capacity of snow is $\nu_w^* \approx 5\%$ (e.g., [1]). For a larger level, liquid water can move inside the snowpack due to gravity, and hydrological runoff begins. Due to the relatively large size of snow crystals, hydraulic conductivity is good, and at $\nu_w > \nu_w^*$, meltwater is filtrated down through the underlaying snow layers. This results into the development of a slush layer at the bottom of the snowpack, where in water-soaked snow ν_w~50%. Measurements of liquid water content have been made in field campaigns by calorimetry [10] and from relative permittivity [11] but it remains a major issue in the field measurement technology [12].

4.1.3 Mechanical Properties

The mechanical properties of snow depend on the density and microstructure, i.e., grain type and size, and their packing and bonding [13–15]. Due to the wide range of snow cover structure and density, the mechanical properties range by orders of magnitude. The microstructure may change notably on short timescales due to snow metamorphosis. The main questions in snow mechanics are the bearing capacity of snow cover and the release of avalanches.

Snow has been modelled in general as a viscoelastic medium (Fig. 4.3). At strain rates greater than 10^{-4}–10^{-3} s^{-1}, snow behaves in nearly linear-elastic manner [16]. The brittle strength of snow is related to density, but there is variability depending on the microstructure, particularly due to the bonds between snow grains. The compressive strength of seasonal snow is 10^3–10^4 Pa ($\approx$ 10^2–10^3 kg-force m^{-2}) that is three orders of magnitude lower than the compressive strength of ice (Fig. 4.4). Snow undergoes viscous flow at low stresses driven by the gravity (e.g., [17, 18]). The flow largely depends on the temperature, density, liquid water content, and microstructure in the snow, and the resulting deformation shows up as densification. Creep rates vary by several orders of magnitude. According to Arenson et al. [13], an overburden stress is needed to reach snow density greater than 350 kg m^{-3}. In the presence of liquid water, the dynamics of snowpack is even more complicated, including the interaction between water flow and microstructure (e.g., [19]). Bonds between grains weaken and the snowpack loses its integrity.

The bearing capacity of snow is connected to methods of moving on snow cover, e.g., by skis, sledges, and feet. As an extreme, in Finland, people have even driven by a small car on snow

Fig. 4.3 Viscoelastic deformation of snow. (Moist snow creeps and may refreeze into interesting, solid geometric forms, *Photograph* Matti Leppäranta)

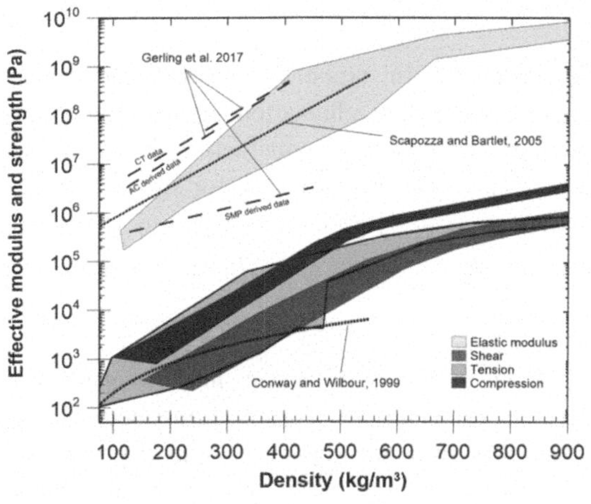

Fig. 4.4 Elastic modulus (upper plot), and shear strength, tensile strength, and compressive strength (lower plots) for snow as a function of snow density. (Arenson et al. [13], based on Mellor [21], and Shapiro et al. [17]. © Elsevier. Printed by permission)

with a crust layer on top, rather for fun. At the strength of 10 kPa or density greater than 300 kg m^{-3} (Fig. 4.4), the strength is about what is needed for a person to stand by two feet on snow. Sámi people have snow terminology related to the conditions under which herders and reindeers can move on snow [20]. The bearing capacity can be mapped with a pressure sensor on snow surface. By this way, a skier can easily map the spatial distribution of the quality of a snow cover.

4.1.4 Thermodynamics of Snow Cover

Thermodynamics of snow is examined for the heat transfer in snowpack and for snow melting which is a major hydrological issue. The thermal properties of snow differ from the thermal properties of ice due to the large fraction of air in snowpack (Table 4.2). Heat capacity is proportional to snow density ρ_s, $(\rho c)_s = \rho_s c_i$, where c_i is the specific heat of ice, and thermal conductivity shows a quadratic law $k_s \propto \rho_s^2$ [22] and can get very low, likewise very good industrial insulators. A widely used formula is by Yen [6]:

$$k_s = 1.889 \cdot \left(\frac{\rho_s}{\rho_i}\right)^{1.885} \frac{\mathrm{W}}{\mathrm{m}\,^\circ\mathrm{C}} \qquad (4.1)$$

The conductivity of snow is one order of magnitude lower than the conductivity of ice. For $\rho_s = 250$ kg m^{-3}, Eq.(4.1) gives $k_s = 0.16$ W m^{-1} °C^{-1} while for ice $k_i \approx 2$ W m^{-1} °C^{-1}. The heat diffusion coefficient of snow is then $D_s = \frac{k_s}{\rho_s c_i} = 0.31 \cdot 10^{-6}$ m^2 s^{-1}, which gives the daily diffusion length scale of $\sqrt{D_s t/\pi} = 9.2$ cm (this length is 31 cm for ice). In general, with Eq. (4.1), the ratio of snow and ice diffusion coefficients is $D_s/D_i \propto (\rho_s/\rho_i)^{0.885}$.

Table 4.2 Thermal conductivity and apparent optical properties of seasonal snow. Ice values are shown for comparison

Property	Unit	Snow	Ice (0 °C)	Comments (snow)
Density	kg m^{-3}	100–400	917	Depends on the weather history
Specific heat	kJ kg^{-1} °C^{-1}	2.1	2.1	2.0 at −15 °C
Thermal conductivity	W m^{-1} °C^{-1}	0.03–0.4	2.1	Proportional to density squared
Heat capacity	kJ m^{-3} °C^{-1}	200–800	1900	Density × specific heat of ice
Diffusion coefficient	10^{-6} m^2 s^{-1}	0.15–0.5	1.1	Thermal conductivity/heat capacity
Albedo	1	0.5–0.9	0.2–0.5	Depend on liquid water content, grain size, cloudiness, solar altitude
Light att. Coefficient	m^{-1}	5–20	0.3–3	

Snow is an optically highly scattering medium where air pockets between snow grains act as the scattering centres. Since these pockets are much larger than the wavelength of light, the scattering spectrum is even. Furthermore, scattering dominates over absorption, particularly in dry snow, so that in normal sunlight snowpack is white. During the polar dim the sun is a little below horizon and the spectrum is dominated by scattered blue light resulting in blueish snow cover. The presence of a certain algae in snow gives red colouring, resulting in so-called pink snow or watermelon snow, and the atmospheric transport of fine sand particles from the Sahara Desert to North Europe results in brown-coloured snow.

The key apparent optical properties of snow are the albedo and the light attenuation coefficient (Table 4.2). They are highly sensitive to the liquid water content of snow and to a lesser degree to the grain size and density, and they are also influenced by cloudiness and solar altitude which determine the directional distribution of the incident irradiance. Dry snow albedo is about 0.9 while the albedo of bare ground is 0.1–0.4, so that a snowack makes a big difference to the surface heat balance. With increasing liquid water content, the albedo decreases, especially at long wavelengths, and at the water holding level the albedo of snow is about 0.5. Grain size is negatively correlated with albedo due scattering effects, with values above 0.9 measured over Antarctic dry, fine-grain snow. The light attenuation coefficient is one order of magnitude larger in snow than in ice, $\kappa_s \sim 10$ m^{-1}. Across a layer of $3\kappa_s^{-1} \sim 0.3$ m, the attenuation of irradiance is 95%. In dry snow, attenuation coefficient is higher for smaller grains due to scattering, and liquid water content reduces scattering thereby improving the light absorption and transfer. In general, the impact of snow properties on the light attenuation coefficient is within a factor of 0.5–2.

Example 4.1 In aquatic sciences, the depth where the irradiance has attenuated to 1% of the incident irradiance is considered the euphotic depth H_e. Using this principle for snowpack, we have $H_e = \frac{1}{\kappa_s} \log[100(1-\alpha)]$, equal to 12 cm for $\alpha = 0.9$, $\kappa_s = 20$ m^{-1} and 78 cm for $\alpha = 0.5$, $\kappa_s = 5$ m^{-1}.

There is a lot of space in snowpack for impurities accumulating from atmospheric deposition. A significant impact on optical properties of snow is due to deposition of light absorbing particles, such as dust and black carbon that lower the albedo and increase light absorption. This phenomenon has been applied locally to melt snow fast for practical reasons, e.g., by spreading ash over snow. Foreign particles in snowpack tend to migrate downward, when possible, by gravity and snow metamorphic processes. In the melting period, meltwater flows down, and impurities become dissolved or suspended in the melt water. They may flow out with the meltwater, but suspended particles also get stuck and concentrate in the snow that shows up as dirty snow patches.

4.2 Mass Balance of Snow

4.2.1 Snow Mass Conservation Law

The hydrological snow issues are to monitor the accumulation of the water storage in snowpack and to predict snow melting and melt water runoff [1]. The water storage is given by the *snow water equivalent (SWE)*, h_{we}, defined as

$$h_{we} = \frac{\rho_s}{\rho_w} h_s \tag{4.2}$$

where $\rho_w = 1000$ kg m^{-3} is the reference density of water and h_s is the snow depth. SWE corresponds to the depth of the potential meltwater layer and is usually expressed in millimetres (as is precipitation). A fraction of snow cover is sublimated and does not add to the runoff.

The mass balance of snow can be written in terms of SWE as

$$\frac{d}{dt} h_{we} = P - E - R + Y \tag{4.3}$$

where P is the precipitation, E is the sublimation/evaporation, R is the runoff, and Y represents the wind drift of snow. Equilibrium exists when precipitation is removed by runoff or snow drift, or in case of very low precipitation, by evaporation/sublimation. Snowstorms and snow accumulation in general bring needs for snow removal. This is a major economic issue for cities and traffic maintenance.

Fig. 4.5 Snow water equivalent measured in the field. The cylinder contains the volume of the vertical sample, and the balance gives the weight in millimetres of liquid water per area (Snow Water Equivalent SWE). (*Photograph* Esko Kuusisto. Printed by permission)

4.2.2 Mass Balance Components

Hydrological snow monitoring systems report the snow depth and water equivalent. The data collection has been traditionally based on snow depth and snow density measurements along snow stake lines of a kilometre length scale (e.g., [23]). Snow density is measured directly from the mass/volume ratio (Fig. 4.5). Airborne remote sensing of the emitted gamma radiation from the ground has provided good SWE data [24]. In satellite remote sensing, passive microwave method has shown feasibility for SWE mapping (e.g., [25]), and radar techniques, especially X-band and Ku-band synthetic aperture radar (SAR), have recently shown applicability for SWE monitoring [26]. Snow cover extent comes out well due to the high contrast between snow and bare ground, but the snow depth and SWE data have lower quality, and snow stakes are still needed for control.

Snow accumulation on land depends on the topography, including the elevation, the slope and orientation of terrain, and surface roughness. Deposition is likely to take place on the windward side of a hill or a building, since near the stagnation point the wind velocity decreases, while on the lee side, turbulent wakes limit the snow deposition. Therefore, at windward forest edge, heavy snow accumulation takes place. Due to interception, there is less accumulation in forest ground than in open fields.

Drifting snow largely modifies the snow accumulation on the ground by transporting snow

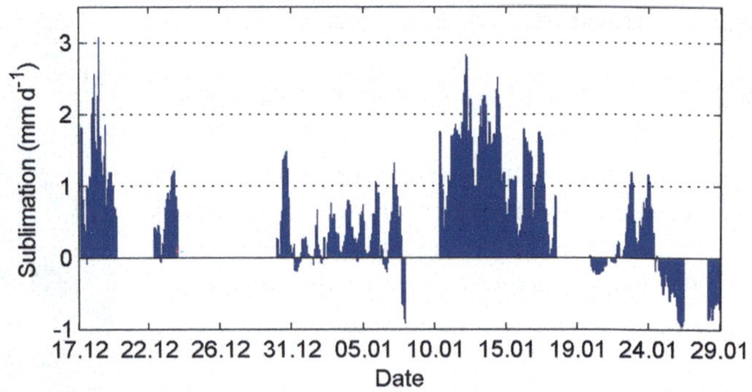

Fig. 4.6 Sublimation in December 2004–January 2005 in a nunataq snow line in Queen Maud Land, Antarctica based on snow measurements, atmospheric surface layer data, and the estimated surface temperature. The *x*-axis numbers show day.month (Leppäranta et al. [27])

during and after snowstorms. When the surface snow layer is loose, turbulent wind can lift snow grains off to be deposited at weak wind spots, resulting in redistribution of the snow on ground. E.g., an uphill wind may transport snow over a mountain peak. Surface roughness influences the turbulence and lifting of snow from the ground, and therefore snow accumulation follows the vegetation. Wind also creates moving snow dunes in similar manner as is seen with sand dunes in deserts.

Sublimation/evaporation can be measured directly in a field campaign but in practice not by an automated system. Direct measurements data are quite limited, and often sublimation is assumed to be small due to the low saturation water vapour pressure in a cold climate. To sublimate a 1 mm solid ice layer in 1 day means a heat loss by 30 W m^{-2}. In an Antarctic nunataq, the mean sublimation from snow was estimated as 0.5 mm ice equivalent per day during summer based on atmospheric surface layer data and snow line monitoring (Fig. 4.6).

With routine data, the aerodynamic method and the energy method can be applied to obtain the latent heat transfer $Q_E = -\rho_w L_E E$ in SWE as is common in hydrology (see Sect. 3.3.3). The former method uses a turbulence model for the latent heat flux while the latter takes it as the residual from the surface heat balance. The turbulent bulk formula gives

$$E = -\frac{\rho_a}{\rho_w} C_E [R q_s(T_a) - q_s(T_0)] U_a \quad (4.4)$$

where q_s is the saturation specific humidity, R is the relative humidity of air, C_E is the bulk transfer coefficient, and U_a is the wind speed. For example, for $R = 50\ \%$, $U_a = 10$ m s^{-1}, and $T_a = T_0 = -5\ °C$, we have $q_s = 0.24\%$ and $E = 1.6$ mm d^{-1} SWE.

4.3 Snow Metamorphosis

4.3.1 Classification of Metamorphoses

The total amount of snow develops according to the mass balance (Eq. 4.3), while the internal structure of snowpack is continuously changing due to mechanical and thermal processes. Snow is characterized by the density, temperature, liquid water content, and grain size and shape, and its evolution, *snow metamorphosis*, is determined by mechanical stress, heat conduction (Eq. 2.13), solar radiation, transport of water vapour and liquid water, and boundary conditions. Water vapour in snowpack is at the saturation vapour pressure which increases with temperature (Fig. 3.1).

Snow metamorphoses are classified into the following categories: (1) Mechanical metamorphosis, (2) Sintering, (3) Sublimation–deposition, and (4) Melt-freeze metamorphosis. Metamorphosis is a continuous process starting from the instant of snow deposition and lasting until the final melting. In general, the mean density increases, and snow crystals become rounded, but the stratification may show up as weak and hard layers as well as recrystallized grains (Fig. 4.7). The stratification and properties of snowpack are examined in a snow pit (Fig. 4.2)

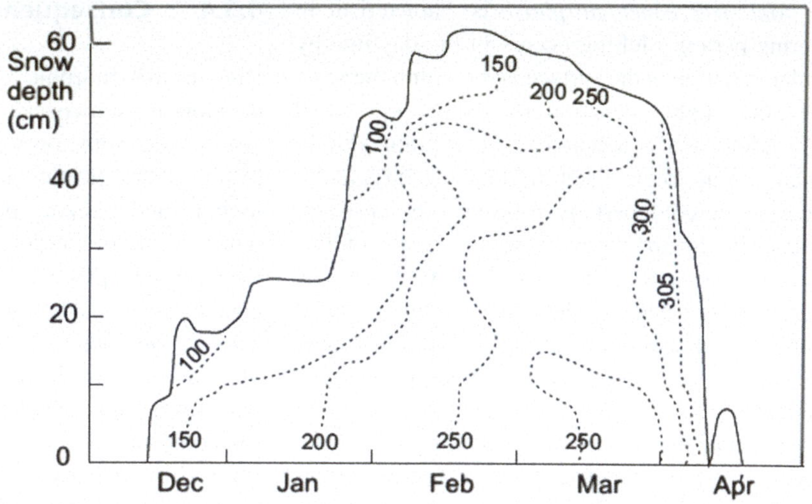

Fig. 4.7 Snow depth (*y*-axis) and vertical snow density profile (contours, in kg m^{-3}) evolution in a site in the boreal zone, South Finland in winter 1968–1969 [28]

and the observations follow a general protocol [4].

4.3.2 Mechanical Metamorphosis and Sintering

Mechanical metamorphosis is due to the wind stress and weight of the snow overload. At the snow surface wind-driven erosion rounds snow grains and packs them denser, while the load of overlaying snow gives a compressive stress $\sigma(h) = \int_{-h}^{0} g\rho_s dz$ which drives the creep of snowpack to denser packing.

In the *sintering* process bonds are formed between snow grains when water vapour is deposited at their contact points, forming necks. Since the water vapour pressure is different at convex and concave surfaces, sharp contacts between grains bridge easily together. Sintering strengthens the snowpack and is considered as *constructive metamorphosis*. Usually in a time scale of days after a snowfall, as can be observed, snow grains get more rounded and sintered together. The density and strength of snow increase, and wind cannot any more lift grains up, also snow shovelling needs more force. Compaction of snow under, e.g., boots, starts up fast sintering when grains are put closer together.

4.3.3 Sublimation–Deposition and Melt–Freeze Metamorphoses

Sublimation–deposition metamorphosis takes place in the presence of a temperature gradient in the snow. Temperature cycles in snowpack are exponentially damped by the *e*-folding depth of $d = \sqrt{D/\pi\omega}$, where ω is the frequency (see Eq. 2.15); e.g., for the daily cycle, $d \approx 10$ cm, so that warm or cold layers may be conserved for long period in deep snow. When air moves in snowpack by convection or diffusion to a warmer (colder) ambient temperature, sublimation (deposition) takes place. Thus, net transport of snow takes place from cold to warm layers, i.e., against the temperature gradient. An important result of this process is the formation of a *depth hoar* layer near the ground–snow interface in a cold period. Near the ground the snow is warmer and loses crystals by sublimation, while ties between grains become stronger due to deposition in the cold upper layers. The time scale of the development is weeks, and the layers change remarkably in their structure. Depth hoar is often the weakest layer in a snowpack and acts in the background of, e.g., triggering cold period avalanches. The sublimation–deposition process is also called *destructive metamorphosis*.

Melt–freeze metamorphosis is characteristic in spring period. Melting occurs in the daytime by solar radiation in the surface layer within the scale of the light penetration depth $\kappa_s^{-1} \sim 10$ cm. Meltwater is adhered to snow grains due to adhesion up to the water holding capacity, and it freezes in cold periods forming ice lenses or crusts in the top layer. The time scale of this process can be from hours to days. In favourable weather, the daily cycles may continue over a long period, and the resulting surface crust layer obtains a high bearing capacity, good for walking and skiing.

Later in the season, when melting dominates over freezing, contacts between snow grains weaken, and the snowpack loses its integrity. Wet snow is different from dry snow as a substance. At a liquid water content of $\nu_w < 8\%$ (pendular regime), pores are filled with air and bonds still exist between grains, and at high liquid water content of $\nu_w > 8\ \%$ (funicular regime), water fills most pores, and the snow is changed into slush. At $\nu_w \sim 15\%$, the snowpack is flooded. Melt water runoff is filtered to ground water storage or flows into lakes and rivers.

4.3.4 Consequences

Due to metamorphic processes, vertical layers develop in snowpack. Each snowfall brings a new layer with own history therefrom. The resulting stratification is especially strong when melting and freezing periods alternate during a winter. The properties of snow change in the metamorphic processes, that has an impact on, e.g., thermal insulation and bearing capacity of snow. Snow shelters for hunters and trekkers exist of two types. In the boreal zone, quinzhees are built from cold, soft snow with artificial bonding forced by pressure (Fig. 4.8), while Inuit can cut 'bricks' from hard tundra snow and wall them together by frictional heating with the cutting knife.

In many applications, the density of snow is a key parameter to estimate snow properties like thermal conductivity and compressive strength. The evolution of density depends primarily on snow load, wind speed, temperature, and solar radiation. Density changes are reflected in snow depth that is accounted for together with SWE in snow monitoring with snow stakes (e.g., [27]).

Fig. 4.8 Quinzhee, a snow shelter, can be constructed by shovelling for a heap and pressuring that by shovels. Constructive metamorphosis will increase the snow strength, and after 2–3 h an overnight shelter can be dug into the heap. Quinzhees have been used since the stone age in the boreal northern Eurasia and America. (*Photograph* Mika Kalakoski)

For snow ecology, the critical snow characteristics are the bearing capacity, and the vertical distribution of density, temperature and sunlight. Snow quality influences moving on and inside the snow, protection from predators, gas transfer, and nourishment of biota. A plant or an animal may be chionophile (snow adapted) or chionophobic (snow avoiding). Beneath a snow cover thicker than $3\kappa_s^{-1} \sim 30$ cm it is dark, and temperature and humidity are more stable and usually higher than on the surface. Snow also damps sounds so that it is quiet beneath snowpack, and a hiker can feel the quietness in a snowy forest. Also, carbon dioxide level can be high. Living is possible on, in or below a snowpack, and animals can move on snow or by digging inside the snow. Plants may even change snow conditions to be more suitable for them.

Sturm & Liston [29] have classified seasonal snow regions to the following categories: ice (snow on top of glaciers and ice sheets), tundra, boreal forest, mountain forest, prairie, maritime, and ephemeral. The classification is based on the topography, vegetation, precipitation, and average temperature with occurrence of warm periods. Tundra snow is hard and densely packed due to the open landscape, winds, and cold climate. Metamorphic processes are more powerful in open area with more snow drift, wind packing, melt–freeze phenomena, and more depth hoar. Thus, forest and open field (prairie, tundra) snowpacks are different: forest is warmer, less windy, and darker than open fields. Maritime snow exists in mid-latitude marine climate with high moisture content and heavy seasonal snow accumulation.

4.4 Energy Balance

4.4.1 Temperature Distribution

Heat conduction in snow cover follows the Fourier's law (Sect. 2.3.2). The vertical heat transfer is written

$$\frac{\partial}{\partial t}(\rho_s c_i T) = \frac{\partial}{\partial z}\left(k_s \frac{\partial T}{\partial z} + Q_{s+} e^{\kappa_s z}\right) \tag{4.5}$$

where z is the vertical coordinate positive up and zero at the snow surface and Q_{s+} is the solar radiation penetrating through the near-surface layer. The boundary conditions are temperatures or fluxes as functions of time. The surface heat flux was discussed in Sect. 3.5. At the bottom boundary, the upcoming heat is used for phase changes at the bottom of the snowpack or conducted up into the snow. At $H \approx 10$ m below the ground surface, the temperature is at the annual average air temperature, and for the snowpack, ground is a cold source on permafrost and a heat source in boreal forest.

In general, density and thermal conductivity (Eq. 4.1) of snow depend on the depth and time. To illustrate the solution, we apply the thermal conduction equation with a fixed density (see Sect. 2.3.2). With $\rho_s = 250$ kg m^{-3}, the diffusion length scale is $\sqrt{D_s t/\pi} = 9.2$ cm for the daily cycle, about the same as the penetration depth of sunlight in snow (~10 cm). In the surface energy balance, the snow albedo is very high (Table 4.2), and about half of the net radiation (infrared part) is absorbed in the near-surface layer. With snow packing, density increases and depth decreases, and therefore the insulation strength of snow cover lowers fast. Due to the low thermal conductivity, high albedo, and high light attenuation, the snow layer protects the ground from weather variations and sunlight, and it has a large impact on the freezing of the ground.

Example 4.2 Consider the equilibrium solution of the heat conduction Eq. (4.5). Clearly,
$k_s \frac{dT}{dz} + Q_{s+} e^{\kappa_s z} = C_1$ (= constant)

Taking the flux surface boundary condition $k_s \frac{dT}{dz}\Big|_{z=0} = Q_0$, we have $C_1 = Q_0 + Q_{s+}$, and by integration, we have

$$T = \frac{Q_0 + Q_{s+}}{k_s} z - \frac{Q_{s+}}{\kappa_s k_s} e^{\kappa_s z} + C_2$$

The first term on the right-hand side gives a linear profile, while the second term gives an exponential correction which vanishes at $-z \gg \kappa_s^{-1}$. Temperature maximum is at the depth

$$z_M = \frac{1}{\kappa_s} \log \frac{Q_0 + Q_{s+}}{Q_{s+}}$$

For a solution, the argument of logarithm must be positive, and for the solution to be below the surface ($z_M < 0$), we must have $-Q_{s+} < Q_0 < 0$. That is, the temperature maximum is below the surface, when the surface balance is negative but the full balance is positive. This is called the cold skin phenomenon, often seen in snow melting starting below the surface. The equilibrium is reached by diffusion with the time scale of z_M^2/D_s. For $z_M \sim \kappa_s^{-1}$, the timescale to reach equilibrium is several hours so that the maximum can develop in a cold and sunny day in spring as has been observed.

4.4.2 Heat Content

The heat content of snow cover consists of sensible and latent heat. It is convenient to take liquid water at the freezing point temperature T_f for the reference level. The heat content, above the reference, of snowpack of depth h per unit surface area is then written as

$$\delta\mathscr{H} = \int_0^h \rho_s \left[c_i(T - T_f) - (1 - \nu_w)L_f\right] dz \quad (4.6)$$

Clearly $\delta\mathscr{H} \leq 0$. In cold snow $\nu_w = 0$ and $T < T_f$, while at the melting stage, $T = T_f$ and $\nu_w \geq 0$. The counter reading of the heat content tells how much heat is needed to turn the whole snowpack to a liquid phase.

The heat content is governed by the latent heat. For example, for $T = -15$ °C, the sensible heat term is 9% of the latent heat term in Eq. (4.6). The negative sensible heat term

$$F_d = -\int_0^h \rho_s c_i (T - T_f) dz \quad (4.7)$$

has been called the *cold content of snow*. It tells how much heat is needed to warm up the snow to the melting point temperature. The *thermal quality of snow* is

$$\theta = \frac{F_d + (1 - \nu_w)F_m}{F_m} \quad (4.8)$$

where $F_m = \rho_s L_f h$ is the latent heat of a hypothetical solid phase snowpack. Clearly $\theta > 1$ ($\theta < 1$) for cold (melting) snowpack.

The evolution of the heat content is driven by external fluxes at the boundaries and solar radiation (see Fig. 3.9):

$$\frac{d}{dt}\delta\mathscr{H} = Q_0 + Q_{s+}\left(1 - e^{-\tilde{\kappa}_s h}\right) + Q_g \quad (4.9)$$

where $\tilde{\kappa}_s$ is the mean light attenuation coefficient in snow, and Q_g is the heat flux from the ground. The fluxes are defined positive into the snow. Snow monitoring (Eq. 4.6) and flux measurements (Eq. 4.9) are independent ways to determine the evolution of the heat content that provides a good control method. The atmospheric and solar fluxes can be obtained from weather data, and the heat flux from ground remains often an unknown but small factor.

Taking the linearized surface heat flux (see Sect. 3.5.3), the conduction equilibrium and the solution are:

$$Q_g = -k_s \frac{T_0 - T_b}{h} = -[K_0 + K_1(T_a - T_0)] \quad (4.10a)$$

$$T_0 = T_a + \frac{K_0 + Q_g}{K_1}, \quad T_b = T_a + \frac{K_0 + Q_g}{K_1} + \frac{Q_g}{k_s/h} \quad (4.10b)$$

where T_b is the temperature on ground, and, for simplicity, the whole solar net flux is included in K_0. For $Q_g > 0$, we have $T_0 < T_b$ and heat is transported from the ground to the atmosphere, and $sgn(T_0 - T_a) =$ sgn $(K_0 + Q_g)$. Only when K_0 is small and $K_1 \sim k_s/h$, the ground temperature signal can be identified in the surface temperature. The solution is valid for T_b, $T_0 < T_f$, otherwise the possibility of melting needs to be considered. For example, let $T_0 = T_f$. Then $T_0 - T_b > 0$ and Eq. 4.10a, 4.10b may hold only if $Q_g < 0$ and heat is transferred from snow surface to ground; else $T_0 = T_b$ and melting takes place at the bottom by the rate of $Q_g/\rho_w L_f$ SWE.

Example 4.3 Take $Q_g = -k_g \frac{T_b - T_g}{\Delta H}$, typically $k_g \sim 1$ W m^{-1} °C^{-1}. In a boreal winter, $T_b - T_g \sim -1$ °C over $\Delta H = 1$ m, and therefore $Q_g \sim 1$ W m^{-2}. For $h \sim 0.5$ m, we have $T_b = T_0 + 2.5$ °C (when T_b, $T_0 < T_f$). In middle winter K_0 is well below zero due to long-wave radiation losses, and $K_0 + Q_g < 0$. Thus $T_0 < T_a$, T_b, i.e., the surface temperature snow is the minimum in the system. Since usually $K_1 > \frac{k_s}{h}$ for $h \sim 0.5$ m, we have $T_a - T_0 < T_b - T_0$.

4.4.3 Melting of Snow

The thermal quality of snowpack (θ) decreases during the melting season. When the cold content has disappeared, $\theta \leq 1$, and at the end $\theta = 0$. Melting usually begins in the surface layer to the depth near the light penetration depth. The temperature of melting snow is at the melting point, but during sub-surface melting due to the cold skin phenomenon the surface temperature may be slightly below. When melting occurs at the surface, the surface heat balance is easy to solve since the surface temperature and humidity are fixed. Since snowpack may experience melt–freeze metamorphosis, it is convenient to treat the snow in terms of SWE. Meltwater is drained for runoff, apart from a small fraction lost by evaporation. Melting takes place earlier in open areas than in forests.

The key factor in melting is the radiation balance. Solar radiation drives the melting opposed by terrestrial radiation losses. At the melting point temperature, the net terrestrial radiation is $(\varepsilon_a - \varepsilon_0)\sigma T_f^4 \approx -50$ W m^{-2} (± 50%) depending on the atmospheric emissivity ε_a (see Sect. 3.2.2). To compensate that, solar radiation over dry snow with albedo of 0.9 needs to be greater than 500 W m^{-2}. Due to metamorphosis of snow surface layer by solar radiation at noon, the albedo may decrease, and once liquid water inclusions form, the positive feedback decreases the albedo further fast. Other possibilities to start melting are liquid precipitation and increased heat gain due to warm air advection where the air–snow temperature difference can be large because the snow surface temperature is at most at the melting point. Heat flux from the ground may be important in a thin snowpack situation due to solar heating of the ground. But, for example, in Antarctica atmospheric heating is limited, and snow melting is almost nothing (e.g., [30]).

Snow melting is obtained from the heat budget Eq. (4.9). Using the linearized heat balance (Sect. 3.5.3), the melting rate of snowpack can be written is in terms of SWE as

$$\frac{dh_{we}}{dt} = -\max\left\{\frac{K_0}{\rho_w L_f} + \frac{K_1}{\rho_w L_f}(T_a - T_f) + \frac{Q_g}{\rho_w L_f}, 0\right\} - E \tag{4.11}$$

It is assumed that in the melting period $T_0 = T_f$ and only positive heat fluxes are accounted for. The parameter K_0 includes the net solar radiation and turns from negative to positive close to the beginning of the melting period. The physics suggests for a constant K_1 and K_0 increasing with time in Eq. (4.11), melt rate consist of a term increasing with time and a degree-day term.

In snow hydrology, a simplified, the so-called degree-day formula has been used where daily snow melting is parameterized as

$$\delta h_{swe} = -\max\{A \cdot (T_a - T_f), 0\} \tag{4.12}$$

where $A \sim 0.3$ cm(°C d)$^{-1}$ is a tuning coefficient. In mid-latitudes, K_0 and air temperature are positively correlated during melting, and the contribution of K_0 has been absorbed in the degree-day coefficient in Eq. (4.12). Therefore, in many applications of Eq. (4.12), A is increasing with time (Fig. 4.9). In the polar zone, melting takes place in summer when the incident solar radiation does not vary much but due to the positive albedo feedback, the net solar radiation increases with time.

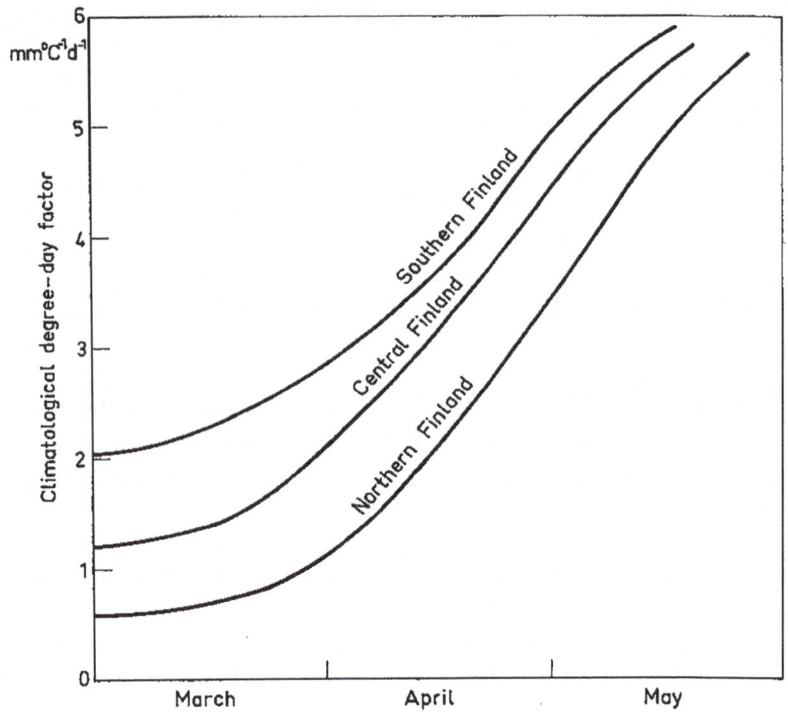

Fig. 4.9 The development of the climatological degree-day factor for snow melting in Finland during the spring months. The strong connection to solar radiation is seen. [23]

4.5 Avalanches

4.5.1 Classification of Avalanches

Snow avalanches are fast movements of snow mass on a mountain slope. They are typical in high mountains with steep slopes, such as the European Alps and the Rocky Mountains in North America. They cause annually natural disasters with many fatalities and large material losses, and their research for protection of settlements and infrastructure as well as for prediction has been an important branch of snow science [31]. The oldest avalanche research laboratory, founded in 1936, is The WSL Institute of Snow and Avalanche Research SLF[4] in Davos, Switzerland.

Avalanches are classified into slab, loose snow, and powder avalanches for their appearance. *Slab avalanche* occurs when a slab of snow slides down as more or less one piece. They contain much snow and are therefore responsible for most of the fatal avalanches. *Loose snow avalanches* are usually small and associated with dry snow. They start in the surface layer in one location and gain more snow and spread wider while moving down. *Powder avalanches* appear like a turbulent flow with very low snow densities in the turbulent eddies. Slides can be dry or wet, depending on the quality of the snow. Powder clouds form when the path is steep, and the snow is dry. For their destructive power, avalanches are divided into five size categories (Table 4.3). For example, a large avalanche has 1000 tons of snow running up to 1 km long path.

4.5.2 Avalanche Mechanics

A snow mass lying on a slope feels the gravity pressing the snow towards the ground and inducing a shear stress which tries to pull the snow down the slope (Fig. 4.10). The snowpack lays stationary if the shear stress τ is less than the shear strength τ_s:

[4] WSL = Forschungsanstalt für Wald, Schnee und Landschaft, SLF = Institut für Schnee- und Lawinenforschung.

Table 4.3 Avalanche size classification with typical physical characteristics [31, 32]

Size	Description	Destructive potential	Mass tons	Path length m	Impact pressure kPa
1	Small	Relatively harmless to people	<10	10–30	1
2	Medium	Could bury, injure, or kill a person	100	50–250	10
3	Large	Could bury a car, destroy a small building (e.g., a wood frame house), or break a few trees	1000	500–1000	100
4	Very large	Could destroy a railway car, large truck, several buildings, or a forest with an area up to 4 ha	10,000	1000–2000	500
5	Extreme	Largest snow avalanches known; could destroy a village or forest of 40 ha	100,000	> 2000	1000

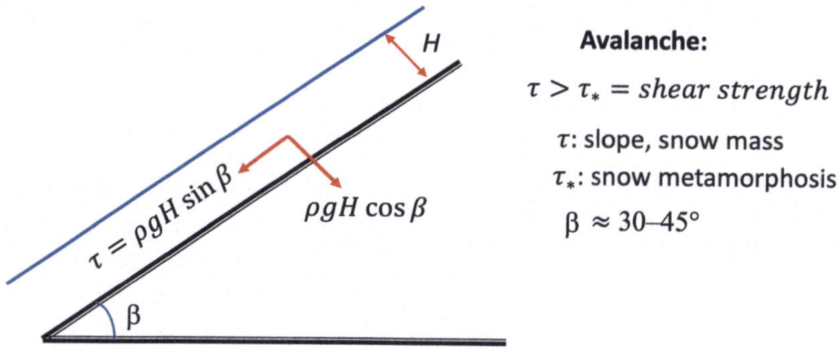

Fig. 4.10 A schematic drawing of an avalanche startup condition for snow pack on a mountain slope. Shear stress increases from snow accumulation and shear strength develops (+/−) in snow metamorphic processes

$$\tau = \rho_s hg \sin\beta < \tau_s \tag{4.13}$$

where β is the slope angle of the terrain. The stress can increase from snow accumulation while the strength can change due to snow metamorphic processes. A snowpack weakens due to depth hoar formation in cold periods and due to loosening of bonds between snow grains in melting periods. Thus, depth hoar may start up an avalanche in mid-winter on a north-facing slope while melt startups are more likely to occur on south-facing slopes. The former case may appear unexpectedly when the weather looks good, and no sign is shown at the surface to an observer.

In practice, the slope angle of the starting zone is 30°–45° in powerful avalanches. In more gentle slopes, snow accumulation is not enough to break the snow, and in steeper slopes there are frequent minor, harmless avalanches. However, in a wet snowpack the slope angle can be little to start. The snow depth must be more than 10–100 cm depending on the slope and roughness of the terrain. A snowpack may experience very slow viscous sliding and deformation that changes the distribution of snow in a terrain, but once an avalanche starts, it is a fast event. The path is divided into three zones: starting zone, track, and runout.

In the track zone, the snow slides down and often tends to follow the topography. The equation of motion shows forcing by gravity and resistance due to friction:

$$\rho_s h \frac{dv}{dt} = \rho_s hg \sin\beta - \mu\rho_s hg \cos\beta \tag{4.14}$$

where v is the snowpack velocity, and μ is the friction coefficient. The static friction coefficient is 0.3–0.6 and the sliding friction coefficient is 0.2–0.5. An avalanche starts when $\tan\beta > \mu_{static}$. When the velocity is more than 10 m s^{-1}, the motion becomes turbulent, and turbulent friction must be added for the resistance.

$$\rho_s R \frac{dv}{dt} = \rho_s Rg \sin\beta - \mu\rho_s Rg \cos\beta - \rho_s C_D v^2 \tag{4.15}$$

where the snow depth has been generalized by the hydraulic radius R to account for the track geometry and C_D is the turbulent drag coefficient. In an open terrain, $R = h$ and the turbulent friction coefficient is C_D~1 [33, 34] depending on the terrain roughness. The steady state solution is

$$v = \sqrt{\frac{\sin\beta - \mu \cos\beta}{C_D} Rg} \tag{4.16}$$

For example, for $\beta = 45°$, $\mu = 0.2$, $C_D = 1$, and $R = 10$ m, we have $v = 7.4 \text{ m s}^{-1}$.

Avalanche velocities are 5–50 m s^{-1} depending on the quality of the snow and the slope angle. In powder snow avalanches, the snow density is of the order of 10 kg m^{-3}, and the velocity can be high. Dry-snow slab avalanches are also fast, whereas wet snow avalanches with snow densities of 300–400 kg m^{-3} are slower, 5–30 m s^{-1}. The path is over an open surface, with typical snow depth of 0.2–2 m, or along a channel where the depth can be much more. In the runout zone, the velocity slows down and the movement stops. The slope is usually less than 15° in the runout zone [35].

4.5.3 Prevention and Prediction

The topography of the terrain is one of the main factors in the avalanche problem. This includes the roughness, slope, and orientation of the terrain. Many slopes are known for recurrent avalanches so that statistical information of past events is also important. Monitoring of snow conditions for the mass and strength of snow provides real-time initial condition which can used in mathematical model simulations, and for forecasting forced by the weather forecast [7].

To prevent future damages, supports have been constructed on slopes to stand for the snow pressure (Fig. 4.11), and road sections have been covered to let the snow slide over. Warning signs have been put to places where major dangers exists. Avalanches kill around 100 people every year in Europe. Particularly the spring period is dangerous due to avalanche favourable weather conditions, and lots of hikers, skiers, and other outdoor people go to risk areas. Potentially fatal avalanches are not easy to predict, and in many

Fig. 4.11 Supports to prevent avalanches on a slope by a road; Hokkaido, Japan. (*Photograph* Matti Leppäranta)

cases, fatal avalanches are triggered by their victims.

Study Problems

1. Consider half-metre deep snow cover with density 200 kg m^{-3} and temperature $-$ 10 °C. How much heat is needed to rise the temperature to the melting point and then to rise the liquid water content to 5%.
2. Air temperature is -10 °C, air humidity is 50%, wind speed is 5 m s^{-1}, and snow surface temperature is -5 °C. How much is sublimation in 1 day? What happens when air temperature changes to 0 °C or $-$ 20 °C with humidity and wind as above?
3. There is an old proverb *New snow is deadly for the old one* in Scandinavia and Finland. How can this be understood in terms of physics?
4. At which depth the amplitude of the daily temperature cycle has damped to 5% of the surface value, when the snow density is a) 100 kg m^{-3}, and b) 300 kg m^{-3}? What is the lag of the temperature cycle in each case?
5. In a boreal site, in April, there was 100-mm liquid precipitation, the average air temperature was 6 °C, wind speed was 5 m s^{-1}, relative humidity was 80%, and solar radiation was 200 W m^{-2}. The snow depth was 60 cm on April 1st. How much was the snow melt and what was the dominant factor?
6. A 4-m-thick slab avalanche occurs on a slope, which has inclination of 30°. What is the speed of avalanche after 1 min from the start and at the steady state? The sliding friction coefficient is 0.4.

References

1. DeWalle D, Rango A (2008) Principles of snow hydrology. Cambridge University Press
2. Male DH (1980) The seasonal snow cover. In: Colbeck S (ed) Dynamics of snow and ice masses. Academic Press, pp 305–395
3. Melack JM, Sadro S, Sickman JO, Dozier J (2021) Lakes and watersheds in the Sierra Nevada of California: responses to environmental change. University of California Press
4. Fierz C, Armstrong RL, Durand Y, Etchevers P, Greene E, McClung DM, Nishimura K, Satyawali PK, Sokratov SA (2009) The international classification for seasonal snow on the ground. In: IHP-VII technical documents in hydrology. IACS contribution (1), vol 83. Unesco-IHP
5. Granberg HB (1998) Snow on sea ice. In: Leppäranta M (ed) Physics of ice-covered seas, vol 2. Helsinki University Press, pp 605–649
6. Yen YC (1981) Review of thermal properties of snow, ice and sea ice. CRREL Report 81-10. U.S. Army Cold Regions Research and Engineering Laboratory, Hanover, NH
7. Bartelt P, Lehning M (2002) A physical SNOWPACK model for the Swiss avalanche warning. Part I: numerical model. Cold Reg Sci Technol 35:123–145
8. Liston GE, Elder K (2006) A distributed snow-evolution modeling system (SnowModel). J Hydrometeorol 7(6):1259–1276
9. Schneebeli M, Sokratov SA (2004) Tomography of temperature gradient metamorphism of snow and associated changes in heat conductivity. Hydrol Process 18:3655–3665
10. Fasani D, Cernuschi F, Colombo LPM (2023) Calorimetric determination of wet snow liquid water content: the effect of test conditions on the calorimeter constant and its impact on the measurement uncertainty. Cold Reg Sci Technol 214:103959
11. Tiuri M, Sihvola A, Nyfors E, Hallikainen M (1984) The complex dielectric constant of snow at microwave frequencies. IEEE J Ocean Eng 9(5):377–382
12. Colbeck S (1978) The difficulties in measuring the water saturation and porosity of snow. J Glaciol 20(82):189–901
13. Arenson L, Colgan W, Marshall HP (2021) Physical, thermal, and mechanical properties of snow, ice, and permafrost. In: Haeberli W, Whiteman C (eds) Snow and ice-related hazards, risks, and disasters, 2nd edn. Elsevier, pp 35–71
14. Mellor M (1977) Engineering properties of snow. J Glaciol 19(81):15–66
15. Salm B (1982) Mechanical properties of snow. Rev Geophys Space Physics 20(1):1–19
16. Marshall HP, Johnson JB (2009) Accurate inversion of high-resolution snow penetrometer signals for microstructural and micromechanical properties. J Geophys Res Earth 114:F04016
17. Shapiro L, Johnson JB, Sturm M, Blaisdell GL (1997) Snow mechanics: review of the state of knowledge and applications. CRREL Tech Rep 97–3:36
18. Yamazaki T, Kondo J, Sakuraoka T, Nakamura T (1993) A one-dimensional model of the evolution of snow-cover characteristics. Ann Glaciol 18:22–26
19. Waldner PA, Schneebeli M, Schultze-Zimmermann U, Flühler H (2004) Effect of snow structure on water flow and solute transport. Hydrol Process 18(7): 1271–1290
20. Gaup Eira IM, Jaedicke C, Magga OH, Maynard NG, Vikhamar-Schuler D, Mathiesen SD (2013)

Traditional Sámi snow terminology and physical snow classification—two ways of knowing. Cold Reg Sci Technol 85:117–130
21. Mellor M (1975) A review of basic snow mechanics. In: The international symposium on snow mechanics, vol 114. IAHS-AISH Publication, Grindelwald, pp 251–291
22. Abel's G (1893) Beobachtungen der täglichen Periode der Temperatur im Schnee und Bestimmung des Wärmeleitungsvermögens des Schnees als Funktion seiner Dichtigkeit. Kaiserliche Akademie der Wissenschaften Repertorium für Meteorologie 16:1–53
23. Kuusisto E (1984) Snow accumulation and snowmelt in Finland. Publications of the water research institute, vol 56. National Board of Waters, p 149
24. Peck EL, Bissell VC, Jones EB, Burge DL (1971) Evaluation of snow water equivalent by airborne measurement of passive terrestrial gamma radiation. Water Resour Res 7(5):1151–1159
25. Dietz AJ, Gessner U, Kuenzer C, Dech S (2011) Remote sensing of snow—a review of available methods. Int J Remote Sens 33(13):4094–4134
26. Tsang L, Durand M, Derksen C, Barros AP, Kang D-H, Lievens H, H-Pe M, Zhu J, Johnson J, King J, Lemmetyinen J, Sandells M, Rutter N, Siqueira P, Nolin A, Osmanoglu B, Vuyovich C, Kim EJ, Taylor D, Merkouriadi I, Brucker L, Navari M, Dumont M, Kelly R, Kim RS, Liao T-H, Xu X (2022) Global monitoring of snow water equivalent using high fequency radar remote sensing. Cryosphere 16:3531–3573
27. Leppäranta M, Järvinen O, Lindgren E (2013) Mass and heat balance of snow patches in Basen nunatak, Dronning Maud land in summer. J Glaciol 59(218): 1093–1105
28. Lemmelä R (1970) On snowmelt, meltwater runoff and groundwater recharge in a glaciofluvial formation in southern Finland. PhLic thesis, Department of Geophysics, University of Helsinki
29. Sturm M, Liston GE (2021) Revisiting the global seasonal snow classification: an updated dataset for earth system applications. J Hydrometeorol 22:2917–2938
30. Järvinen O, Leppäranta M, Vehviläinen J (2013) One-year records from automatic snow stations in Western Dronning Maud land, Antarctica. Antarctic Sci 25(5):711–728
31. Schweizer J, Bartelt P, van Hervijnen A (2021) Snow avalanches. In: Haeberli W, Whiteman C (eds) Snow and ice-related hazards, risks, and disasters, 2nd edn. Elsevier
32. McClung DM, Schaerer P (2006) The avalanche handbook. The Mountaineers Books
33. Kyburz ML, Sovilla B, Gaume J, Ancey C (2022) Physics-based estimates of drag coefficients for the impact pressure calculation of dense snow avalanches. Eng Struct 254:113478
34. Martinelli M, Lang TE, Mears AI (1980) Calculation of avalanche friction coefficients from field data. J Glaciol 26(94):198–119
35. Jamieson B (2001) Snow avalanches. In: Brooks GR (ed) A synthesis of geological hazards in Canada. Bulletin, vol 548. Geological Survey of Canada, pp 81–100

Frozen Ground

5

Melting of permafrost results in forest damage in the boreal zone, Yakutia, in the summer of 2017. (*Photograph Matti Leppäranta)*

This chapter presents the seasonally frozen ground and permafrost. Frozen ground is a practical issue in the seasonal snow and ice zone due to ground deformation during freezing and melting. Scientific research on permafrost started in the nineteenth century in Siberia and, soon after, in North America [1]. In the twenty-first century, Chinese scientists have been very active

M. Leppäranta, P. Uotila, *Geophysics of the Cryosphere*, Springer Textbooks in Earth Sciences, Geography and Environment, https://doi.org/10.1007/978-3-031-93071-3_5

on this topic due to their large mountain permafrost areas, especially in the Tibetan Plateau. Early research was motivated by observations indicating that the temperature increases with depth in the non-frozen ground at temperate latitudes. Permafrost was an exception requiring explanation. Later, various construction and engineering projects in the permafrost region have required advanced knowledge of its properties and evolution.

5.1 Frozen Landforms

5.1.1 Seasonally Frozen Ground and Permafrost

Frozen ground is a part of the Earth's soil layer and lithosphere. It forms from the soil water and groundwater storage when the temperature falls below freezing. *Seasonally frozen ground* persists, by definition, for at least 15 days, and in high latitudes or high mountains, it can survive over several months. When the ground stays frozen for at least 2 years, it is classified as *permafrost*, perennially frozen ground [2, 3]. The mean annual air temperature in permafrost regions is typically less than −2 °C. The top permafrost layer, which melts in summer and re-freezes in winter, is called the *active layer*, and the boundary between the active layer and permanently frozen ground is called the *permafrost table.*

The depth of seasonally frozen ground is of the order of 1 m. For example, in Finland (60–70°N), the frozen ground period is about 5–7 months, and its depth is about 0.5 m in the south and almost 2 m in the north. Permafrost depth can reach hundreds of metres beneath the surface. The oldest known permafrost, found in Siberia, is 650,000 years old [4]. The ratio of ice to the total frozen ground volume varies within tens of per cents. Frozen ground is characterized by its areal extent, vertical temperature, thickness, and active layer thickness (Portrait 5.1).

Portrait 5.1 Academician Pavel Ivanovich Melnikov (1908–1994) was one of the pioneers of the permafrost research. His scientific career spanned over seven decades. He founded the permafrost institute in Yakutsk in 1960, later named Melnikov Permafrost Institute of the Siberian Branch of the Russian Academy of Science. His theoretical and experimental contributions to engineering and geocryology include the development of relationships to predict the physical and mechanical characteristics of permafrost soils, an understanding of the mechanisms by which wedge ice is formed, and the design methods and construction techniques for buildings and structures on permafrost. The photographs shows the bust of P.I. Melnikov in front of the Melnikov Permafrost Institute in Yakutsk. (*Photograph* Matti Leppäranta)

5.1.2 Other Forms of Frozen Land

When the ground freezes, the landscape changes significantly due to various physical processes related to the change from liquid to solid phase of water and the consequent expansion. *Pore ice* can form in tiny pores in the soil. Pore ice holds the soil together and exists at a tiny scale between the soil grains. Water can segregate due to frost heaving to larger ice chunks in the form of layers and lenses as segregation ice. Vein ice forms in narrow, elongated cracks in contracted cold soil when water freezes. Eventually, vertically oriented ice wedges may form if the process is

Fig. 5.1 Needle ice in the ground. Needle ice forms from capillary groundwater in cold weather. (*Photograph* Thomas Bresson © CC BY 2.0)

repeated. The fourth form of ice is intrusive ice, generated when water flows into the frozen ground and freezes. Pressure differences usually force such flows. On the soil surface, needle ice can form as upward-oriented elongated ice crystals growing towards the heat loss direction (Fig. 5.1). The size of these crystals varies from a few millimetres to several centimetres. Needle ice formation can occur anywhere the soil surface freezes.

Frost heaving is related to the freezing of soil when the ice formation lifts the soil. When combined with thawing periods, soil material can be sorted, denoted by cryoturbation. In the sorting process, delicate and lighter material has the advantage of being pushed up by a frost heave at the freezing plane of the active layer. This can result in ground surfaces where sorted and non-sorted areas alternate as polygons and circles. Ground ice also deforms the soil above it and generates various surface landforms by frost heaves. As the temperature decreases, soil contracts and cracks. The size and distribution of these cracks depend on temperature and its rate of decline. *Palsa* is a Finnish word for a small mound of peat, 1–8 m high, rising out of a mire with a cyclic development from a few decades to centuries (Fig. 5.2). It contains a permafrost core of peat and silt, tiny ice crystals, and thin layers of segregated ice that at least partially survive over summer [5]. *Pingos* are bigger frost mounds with a large ice core that can be up to 50 m high and 300 m in diameter. They form and grow by repeated water injections from below that freeze and expand. Both palsas and pingos are perennial features of the surface and survive over the summer season.

Related to permafrost thaw is *thermokarst* terrain. Where ice-rich permafrost melts, the soil subsides, forming depressions, often filled with water. These lakes can be many metres deep, kilometres wide and clustered in the landscape. A supra-permafrost *talik* can form under the lake when a thermokarst lake is deep enough. Thermokarst lakes grow when thawing advances and can eventually drain and disappear.

Fig. 5.2 A palsa, peat mound with permanently frozen core, in Finnish Lapland. Palsas are common in the discontinuous permafrost zone. (*Photograph* Miska Luoto, printed by permission)

Fig. 5.3 Melting permafrost at the coast of Alaska. This kind of shoreline erosion can be strengthened if the ice-free period in summer becomes longer. (*Photograph* Benjamin Jones, USGS/Alaska Science Center)

Frost heave and thaw pose challenges to plants, animals, and humans who need to adapt to the changing soil elevation (Fig. 5.3). This must be considered when constructing roads, buildings, and other regional infrastructure. Building foundations must either be insulated from the ground frost or extended below the frost layer to prevent damage. For example, in Finland, foundation support structures are instructed to extend twice to the depth of the ground frost layer (0.5–2 m). In spring, the meltwater cannot be absorbed to the frozen ground below, and roads can be damaged, become soggy and bumpy, and their pavement can break. The period when the traffic is stopped due to frost heave and thaw problems has a specific name in northern languages, e.g., kelirikko ('broken connections') in Finnish, распу́тица (rasputitsa, 'season of bad roads') in Russian.

5.2 Frozen Ground Temperature

5.2.1 Heat Flow in Ground

The temperature (T) in the ground follows the thermal conduction equation (see Sect. 2.3.2):

$$\rho c \frac{\partial T}{\partial t} = \frac{\partial}{\partial z} k \frac{\partial T}{\partial z} \tag{5.1}$$

where ρ is the density, c is the specific heat, k is the thermal conductivity, t is the time, and z is the vertical coordinate positive up. Representative thermal properties of soil are given in Table 5.1. When the diffusion coefficient $D = k/\rho c$ is taken as a constant, the solution with the sine wave (frequency ω) surface boundary condition equals sine waves delaying with depth with exponentially damping amplitude (see Eq. 2.15). The e-folding depth is $\ell = \sqrt{D/\pi\omega}$ and the delay is half cycle at $z = \pi\ell$; e.g., for annual cycle, $2\pi\omega = 1$ a, we have ℓ = 3.9 m for mineral soil and ℓ = 1.0 m for organic soil.

The energy balance over frozen land and water surfaces is discussed in detail in Sect. 3.5. Here, the precise estimation of surface energy balance of frozen ground is highlighted. To correctly estimate the evolution of the active layer, a proper estimation of the surface energy balance plays a key role. The surface energy balance in cold regions shows large temporal variability through daily to seasonal time scales. For example, on a summer day, the surface receives solar radiation, its moisture evaporates, and the ground warms, inducing net heat flux from the surface to the ground below. The opposite occurs at night when the atmospheric humidity is condensed at the ground surface, net heat flux is directed from the ground to the surface, and the ground loses heat to the night air.

Vegetation and snow cover are related to the insulation from cold winter air. Thicker snow is more effective, for instance, under low shrub vegetation than under a terrain without vegetation. This difference affects whether and at what level the permafrost exists and how deep the active layer penetrates. There may be no permafrost under a thick snow layer, but just a short distance away, where the snow is much thinner, permafrost might exist just a few metres below the ground surface.

The key surface variables related to the energy balance are albedo, thermal conductivity, roughness, and emissivity. They depend on the material properties of the surface, such as the grain structure, porosity, and moisture content, and they vary greatly depending on whether the surface is ice, snow, sand, grass, shrubs, trees, and so on. For example, the emissivity of seasonal snow is 0.96–0.99, while the emissivity of frozen soil is less, about 0.92 [7]. The thermal conductivity of snow varies greatly from less than 0.1 to 0.3 W m^{-1} K^{-1} (e.g., [7], see also Sect. 4.1.4), being proportional to the squared snow density [8]. In comparison, the conductivity of sand is 0.5–4 W m^{-1} °C^{-1}, but in peat lands, it can be as low as 0.03 [8] W m^{-1} K^{-1}. Moreover, various surface types often change spatially at short length scales. For large spatial scales, effective values are determined, for instance, by averaging available small-scale values or deriving

Table 5.1 Thermal properties of soil and granite [6], and water and ice

Medium	Density kg m^{-3}	Specific heat kJ kg^{-1} °C^{-1}	Thermal conductivity W m^{-1} °C^{-1}	Diffusion coefficient m^2 s^{-1}
Mineral soil (20 °C)	2650	0.73	2.9	$1.5 \cdot 10^{-6}$
Organic soil (20 °C)	1300	1.90	0.25	$0.10 \cdot 10^{-6}$
Granite (20 °C)	2640	0.72	3.1	$1.6 \cdot 10^{-6}$
Water (0 °C)	1000	4.22	0.56	$0.13 \cdot 10^{-6}$
Ice (0 °C)	917	2.11	2.14	$1.1 \cdot 10^{-6}$

them from remote observations with a large footprint.

Different ecosystems have different surface characteristics, which affect their weather and climate response. For example, tundra tends to be warmer (colder) in summer (winter) than forest at the same location. On the other hand, depending on the canopy height and density of trees, the snow thickness may be greater or less in the forest than in the tundra. There tends to be less snow under tall trees but more snow, mainly drift snow, at low trees and bushes (see Chap. 4 for snow physics). All these factors affect the temperature distribution and the depth of the permafrost table in the ground.

In addition to the thermal properties of the ground, the hydraulic conductivity, which describes how well water moves through ground pores, decreases several orders of magnitude when the ground temperature is reduced from 0.5 °C to 0 °C for saturated porous media and when the ground gets frozen. Consequently, groundwater flows are strongly affected by permafrost. In extreme cases, saturated, fully frozen ground is impenetrable for groundwater flow, but this is not always true. At the beginning of the freezing process, some flow can be routed through the frozen ground along empty or liquid water pores.

5.2.2 Seasonally Freezing Ground

In autumn, the ground loses heat by longwave radiation and turbulent fluxes. At about 10 m depth, the temperature is at the annual mean air temperature, and in the cold season, there is then an upward heat flux of $Q_g = k\frac{\partial T}{\partial z} \sim 1$ W m^{-2} in the top ground layer. Considering the linearised surface heat balance $Q_0 = K_0 + K_1(T_a - T_0)$ (Eq. 3.34; see Sect. 3.5), where T_0 and T_a are the surface and air temperature, the equilibrium surface temperature is

$$T_0 = T_a + \frac{K_0 + Q_g}{K_1} \tag{5.2}$$

The parameter K_0 depends on the radiation balance, climatologically $|K_0| < 50$ W m^{-2} and $K_1 \approx 10 - 20$ W m^{-2}. In practice, the air temperature needs to be below 1–2 °C in high-latitude fall season for the ground surface to freeze, and in low-latitude mountains, the critical air temperature is a few degrees less. The freezing point temperature is likely to be slightly below 0 °C due to the pore water salinity, but often it is just assumed that $T_f = 0$ °C.

The growth of the seasonal frozen ground can be solved by the heat conducting equation presented in Sect. 2.3.2. Ground is non-transparent to solar radiation that leaves the source term out, but the thermal properties of ground show space-time variability. Snow accumulation produces an insulating layer above ground that is a critical issue in estimating the evolution of the frozen layer. Thus, a proper model is a two-layer (snow and ground) system with atmosphere–snow coupling and weak heat flux from deep soil towards the frozen layer. Chap. 4 discusses the snow thermodynamics.

Quasi-steady approximation provides a way to obtain an analytic solution for the depth of the seasonally frozen layer. This approach can be motivated by the small depth and, consequently, the small thermal inertia of the frozen layer. For a bare ground, the quasi-steady equation reads

$$\rho_i L_f \nu \frac{dh_f}{dt} = k\frac{T_f - T_0}{h_f} = -[K_0 + K_1(T_a - T_0)] > 0 \tag{5.3}$$

where ρ_i is the ice density, L_f is the latent heat of freezing, ν is the water content of the ground, h_f is the thickness of the frozen layer, T_f is the freezing point temperature, and T_0 is the ground surface temperature. In the research of inland surface waters, Eq. (5.3) is credited to Stefan (see Sect. 7.3.1). The solution is, assuming the thermal parameters constant,

$$h_f(t) = \sqrt{(h_{f0} + b)^2 + \frac{a^2}{\nu}S} - b \tag{5.4}$$

$$a^2 = \frac{2k}{\rho_i L_f}, b = \frac{k}{K_1}, S = \int_0^t \max(T_f - T_R - T_a, 0°C)\, d\tau$$

where h_{f0} is the initial depth and $T_R = K_0/K_1$ is a correction term that depends on the zeroth-order heat balance (see Sect. 3.5.3). In winter, roughly speaking $|T_R| < 3\ °C$. This formula gives the maximum potential depth when there is no snow cover.

Example 5.1 Take $a = 9$ cm^2 °C^{-1} d^{-1}, $\nu = 0.5$, $b = 10$ cm, and $S = 1000$ °C d. Then, the frozen layer depth is $h_f = 125$ cm.

To improve the model, a snow layer can be added. Due to the low thermal conductivity of snow, a large temperature gradient exists across snow in cold weather. This poses an issue for predicting frozen ground evolution because air temperature and snow surface temperature measurements are more commonly available than ground surface temperatures. Assuming snow surface temperature is equal to air temperature T_a, we have for the growth:

$$\rho_i L_f \nu \frac{dh_f}{dt} = k\frac{T_f - T_0}{h_f} = k_s\frac{T_0 - T_a}{h_s} \geq 0 \quad (5.5)$$

where k_s is the thermal conductivity of snow, and h_s is the snow depth. This equation is based on the continuity of the heat flux from the bottom of the frozen layer to the air. Eliminating the ground surface temperature, the growth law is

$$\rho_i L_f \nu \frac{dh_f}{dt} = k\frac{T_f - T_a}{h_f + b + (k/k_s)h_s} \quad (5.6)$$

This can be integrated by daily time steps when snow accumulation and air temperature data are available. For a simple illustration, if $h_s = \overline{h}_s =$ constant, we have

$$h_f(t) = \sqrt{\left(h_{f0} + b + \frac{k}{k_s}\overline{h}_s\right)^2 + \frac{a^2}{\nu}S} - \left(b + \frac{k}{k_s}\overline{h}_s\right) \quad (5.7)$$

Example 5.2 Continue Example 5.1. If $\overline{h}_s \sim 50$ cm and $k/k_s \sim 5$, the frozen layer depth is reduced to 33 cm.

Although the surface temperature in winter is closely related to the air temperature above, it can occasionally be quite different. At the extreme, on a calm, clear sky winter night, the thermal radiation loss from the land surface is substantial, and its temperature can be several degrees lower than the air temperature. This can be solved by coupling the surface temperature with the surface heat balance (see Sect. 3.5), but the difference is negligible in terms of the maximum annual depth of the frozen layer.

The geothermal heat flux can be added with the first term in Eq. (5.5) since it must also be conducted to air through ground and snow. Usually, it is a minor factor. However, the geothermal heat flux can prevent ground freezing when snow accumulation is large. We have the steady state condition from Eq. (5.5) as

$$Q_g = k_s\frac{T_0 - T_a}{h_s} \quad (5.8)$$

Taking $T_0 = T_f$, we can solve for snow thickness that represents the minimum for the unfrozen soil condition. For example, the soil is primarily unfrozen in Hokkaido, Japan due to the thick snow cover. The winter mean air temperature in Sapporo, southern Hokkaido, is about −5 °C, and with $Q_g \approx 2$ W m^{-2}, and $k_s \approx 0.2$ W °C^{-1} m^{-2}, we have $h_s \approx 0.5$ m.

Various ground types, such as clay, silt, rocks, and debris, exhibit different thermal and hydraulic properties, such as conductivities. They behave differently during freezing. For example, clay forms ice lenses more easily than other soil types due to its more effective cryostatic suction. This means that in the freezing ground, the increasing pore pressure induces capillary suction that causes an upward flux of water towards smaller pore pressure. A complication is that these various material structures often coexist in the ground. In addition to the limitations of the quasi-steady model to specific ground types that mainly affect the thermal properties of the media, other shortcomings exist. These include the equality between ground and air temperatures, which is rarely true. The thermal properties of the vegetation and the temperature profile in the atmospheric boundary layer should be considered.

5.2.3 Thawing Processes

The frozen ground layer melts primarily from the top surface down. Melting begins when the surface temperature has reached the melting point, and the heat balance remains positive. Snow cover is crucial for surface thaw in spring and summer. Due to its high albedo and insulating capabilities, it delays the heating and melting of the ground underneath. Before the onset of melting, snow must mostly disappear since the light penetration distance is around 10 cm in snow.

The melting front penetrates downward as

$$\rho_i L_f \nu \frac{dh}{dt} = -k\frac{T_0 - T_f}{h} = -[K_0 + K_1(T_a - T_0)] < 0 \quad (5.9)$$

where h is the depth of the frozen-melted soil interface. Taking the initial condition $h(0) = 0$, the solution is

$$h = \sqrt{\frac{a^2}{\nu}P + b^2} - b, P = \int_0^t \max(T_a + T_R - T_f, 0\,°\mathrm{C})d\tau \quad (5.10)$$

where P is the sum of positive-degree-days, and $T_R = K_0/K_1$. This law is precisely the inverse of the growth law (Eq. 5.4). For example, for $T_a = 5\,°\,C$ and $\nu = 0.5$, melting of a 50-cm frozen ground layer takes 28 days. Melting of frozen ground from below is slow, for $Q_g \approx 2$ W m^{-2}, the bottom melt rate is $Q_g/\rho_i L_f \nu$~0.11 cm day^{-1}.

We can combine the growth and melting formula to examine the seasonality of the frozen ground. In the absence of snow cover, we have the seasonality condition:

$$h_{f,\max} - h = \left\{\sqrt{\frac{a^2}{\nu}S + b^2} - b\right\} - \left\{\sqrt{\frac{a^2}{\nu}P + b^2} - b\right\} \leq 0 \quad (5.11)$$

which gives $S \leq P$. Combining Eqs. (5.4) and (5.10), we have the first approximation,

$$\langle T_a \rangle \leq T_f - \langle T_R \rangle \quad (5.12)$$

where $\langle\cdot\rangle$ refers to the annual average. The term $\langle T_R \rangle$ is related to the mean radiation balance (see Sect. 3.5), $\langle T_R \rangle > 0$; e.g., in Finland, $\langle T_R \rangle$~2 ° C [9]. Physically, this means that the solar heating of the atmosphere goes partly via the Earth's surface. Thus, Eq. (5.12) explains the general empirical result that permafrost occurs where the mean annual temperature is less than −2 °C. In general, the margin needs to be determined by observations to account for the ground quality and snow conditions. In Canada, it has been observed that the boundary of discontinuous permafrost lies along the −1 °C isotherm [10].

Adding snow cover works against ground freezing. Typically, snow cover melts an average of 2–3 cm per day during decay. Still, snow accumulation's impact on reducing the winter growth of frozen ground is more substantial. At extremes, heavy snow accumulation can prevent the ground from freezing. Thus, Eq. (5.12) works as an approximation for thin snow conditions. For thick snow, it is clear that the mean annual air temperature must be significantly lower for permafrost conditions.

5.3 Permafrost

5.3.1 Permafrost Growth and Equilibrium

Permafrost refers, by definition, to the frozen ground that is at least 2 years old. Thus, an extreme 1-year fluctuation is excluded. Then, the depth of the frozen ground is enough to be protected from 1 warm year. In an average year, the freezing-degree-days are split into two parts: one needed to freeze the melted summer layer and another to grow permafrost, say, $S = S_1 + \Delta S$. Thus, the thickness of the frozen ground after the second winter is

$$h_{f2} = \sqrt{\left(h_{f1} + b + \frac{k}{k_s}\bar{h}_s\right)^2 + \frac{a^2}{\nu}\Delta S} - \left(b + \frac{k}{k_s}\bar{h}_s\right) \quad (5.13)$$

where h_{f1} is the frozen ground thickness after the first year.

With time, the seasonal cycle of air temperature takes place in the active layer with a depth scale of 1 m. At the bottom of the active layer, the temperature is close to the annual mean local air temperature, and the permafrost layer deepens year by year. The equilibrium depth h_E is reached when the heat flux from the unfrozen ground and heat loss to the atmosphere are in balance:

$$h_E \approx k \frac{T_f - \langle T_a \rangle}{Q_g} \tag{5.14}$$

For the heat flux of $Q_g \sim 50$ mW m^{-2} deep in the ground [11] and $\langle T_a \rangle \sim -5$ °C, we have $h_E \sim 100$ m, which corresponds to the observed scale of the permafrost depth. The time scale of deepening is roughly $h_E^2/h_{f1} \sim 10^4$ years.

At some depth, the amplitude of annual temperature cycle becomes zero, and seasonal variations do not affect the permafrost temperatures below that level. Above this depth is the permafrost table, where the maximum yearly temperature reaches 0 °C and bore water melts in summer. Notably, the soil can contain salts and minerals that depress the freezing point below 0 °C. Then, in the transition zone just below the permafrost table, the temperature remains below 0 °C all year but above the freezing point, and a significant fraction of pore water may stay liquid. The role of snow in delaying the permafrost thaw is particularly strong in the regions of discontinuous permafrost as it reduces the heating of unfrozen ground within the permafrost regions, which then reduces the heating of the permafrost itself.

Numerical models for permafrost evolution apply the vertical heat conduction equation (e.g., [12]) with information on soil types, including their thermal conductivities and heat capacities. In these models, the buffer layer between the atmosphere and the ground consists of vegetation canopy, snow cover and the organic layer between snow and frozen ground. Underneath the permafrost, geothermal heat fluxes are given. Practically, it is nearly constant with a mean of about 50 mW m^{-2} in the continental terrains [11]. This way, the vertical temperature profile through permafrost can be realistically estimated, given that the soil structure and surface are correctly observed.

The most advanced permafrost models are three-dimensional in space. They are used to predict the evolution of permafrost in climate models, where they are coupled with the atmosphere as a part of a land surface model. These models consider the vertical and lateral heat, water, and carbon fluxes. An essential consequence of land-air coupling is that such a configuration can capture the permafrost-carbon feedback where the release of carbon from thawing permafrost as methane and carbon dioxide increases the air temperature due to strengthening the greenhouse effect. Such an increase further accelerates the permafrost thaw and the release of its carbon.

5.3.2 Taliks

Hydrological systems, such as lakes and rivers, effectively delay and prevent the freezing of the surrounding ground and promote its melt [13]. This is due to the exceptionally high heat capacity of water. Taliks, unfrozen parts within the permafrost, are often related to these large water bodies (Fig. 5.4). Taliks are heat sources to permafrost since their temperature is higher than permafrost temperature. In a way, taliks resemble supraglacial lakes or sea ice melt ponds for their geophysics. A closed talik is surrounded by frozen permafrost, while an open talik is connected to other unfrozen ground zones.

Taliks can form between the permafrost table and the frozen part of the active layer as suprapermafrost taliks. Intra-permafrost taliks can exist inside permafrost, while sub-permafrost taliks are found under the cryopeg. Their emergence and growth are often related to shrinking or expanding lakes and rivers, which change the groundwater flow and the permafrost table under the water body so that a talik can extend further down and penetrate the permafrost layer and become an intra-permafrost talik [13].

Groundwater is an effective carrier of heat in non-frozen ground, including taliks. Therefore, taliks associated with water bodies heat and can

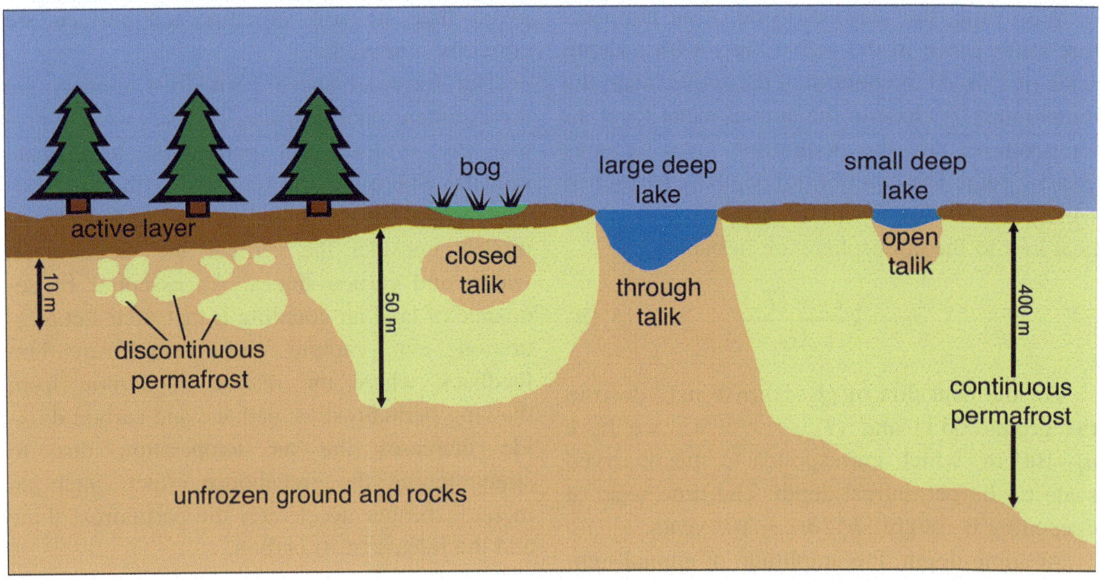

Fig. 5.4 A schematic picture of three types of taliks: open, closed and through talik. Taliks are permanently unfrozen layers in permafrost regions. (*Source* Timonrey/Wikipedia © CC BY-SA 3.0)

melt the adjacent permafrost and are more likely to go from closed to open in a warmer climate extending down to the perennially unfrozen ground. In the process, continuous permafrost areas become discontinuous. A similar explanation can be used to differentiate the existence of continuous and discontinuous permafrost in colder and warmer climates simultaneously but at different locations. In changing permafrost and taliks, the surface and groundwater hydrology also change with the volumes of lakes, rivers, and aquifer flows. This also appears to be a positive feedback in the ground thaw related to the high melt potential of liquid water. In summary, drastic differences exist in permafrost distribution and related hydrology between cold and warm climates.

5.4 Vertical Extent of Seasonally Frozen Ground and Permafrost

5.4.1 Stratification of Permafrost

Seasonally frozen ground consists of the top layer extending up to 1–2 m, and its age is, by definition, less than 1 year. Permafrost is often very old, thousands of years, and may extend to hundreds of metres in depth. The oldest permafrost has been found in Siberia and Yukon, Canada, and is 650,000 years old [4, 14]. However, permafrost is relatively thin compared to the total thickness of the lithosphere, 10–100 km. The deepest permafrost in East Siberia (1650 m) is only about 1% of the lithosphere thickness.

On the top of permafrost, the active layer that thaws every summer, or some summers, can usually be found (Fig. 5.5). The thickness of the active layer scales to 1 m, varying depending on the summer temperature, soil type, and the water content of the ground. It indicates the maximum annual thawing depth, the permafrost table, under which is the perennially cryotic layer. If the active layer reaches the permafrost table, the permafrost underneath is said to be active. Otherwise, a non-frozen layer exists between the permafrost table and the base of the active layer and the permafrost underneath is passive.

In autumn, the freezing of pore water in the active layer releases latent heat, which warms the soil above, causing the ground temperature to remain at 0 °C for a prolonged period. The opposite happens in spring when thawing requires

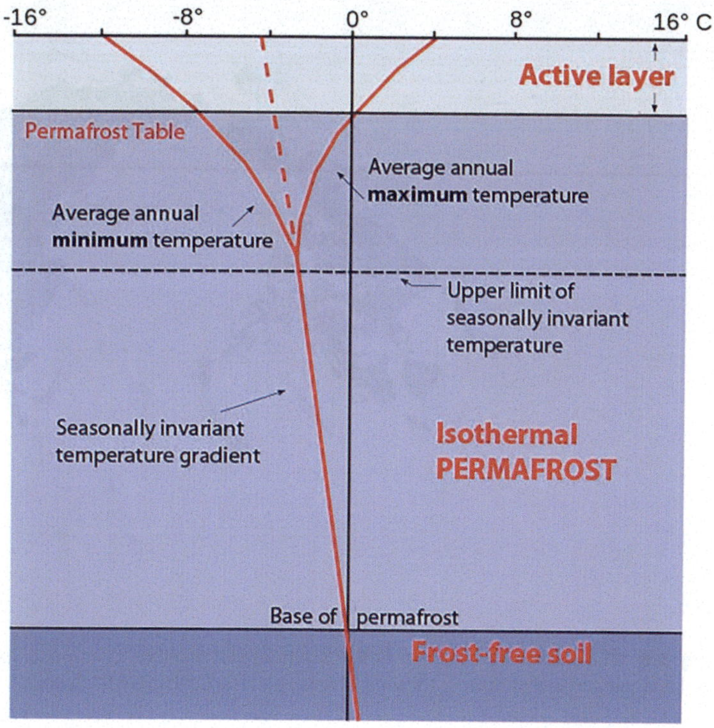

Fig. 5.5 Temperature profile through permafrost. The active layer melts in summer, its lower boundary is the permafrost table. (*Source* German Wikipedia/HylgeriaK, translation by HopsonRoad © Universal Public Domain Dedication)

heat, and the ground temperature does not increase until all ice at that location melts.

In the lower part of permafrost, where temperature increases linearly with depth due to the geothermal heat flux, there is a layer called cryopeg where the temperature is below 0 °C but above the freezing point so that as in the transition zone, a significant amount of pore water remains liquid. The freezing point depression may be due to salinity (0.05 °C per 1 ppt) and pressure (0.7 °C per 100 bar, see Sect. 2.1). Below the cryopeg layer, permafrost ends, and the ground remains perennially unfrozen. The zero curtain is an exciting effect that delays cooling in autumn and warming in spring.

5.4.2 Geographical Extent of Frozen Ground and Permafrost

In the Northern Hemisphere, about 55–60% of the land surface gets at least seasonally frozen. Permafrost covers 11% of Earth's exposed land surface, and it is more extensive in the Northern Hemisphere, where it covers 22% of the land area. Thus, most permafrost is located in the Arctic zone of the Northern Hemisphere (Fig. 5.6). The location of frozen ground is primarily determined by air temperature and snow cover. Therefore, its large-scale distribution follows the roughly zonal climatological isotherms. However, continental climates vary considerably, and, for example, in colder Eurasia, permafrost extends further south, underlying about 65% of the landscape, while in North America, where it underlies about 35% of the landscape. To limit or prevent melting and allow freezing in winter, temperatures should then stay on average below 0 °C. In the Northern Hemisphere, continuous permafrost regions are approximately located north of −8 °C isotherm and seasonally frozen ground north of −1 °C isotherm [15]. Naturally, as these isotherms move along with changing climates, frozen ground regions also experience transitions.

Continuous permafrost that underlies 90–100% of the landscape exists across large areas in Siberia, northern Canada, including its

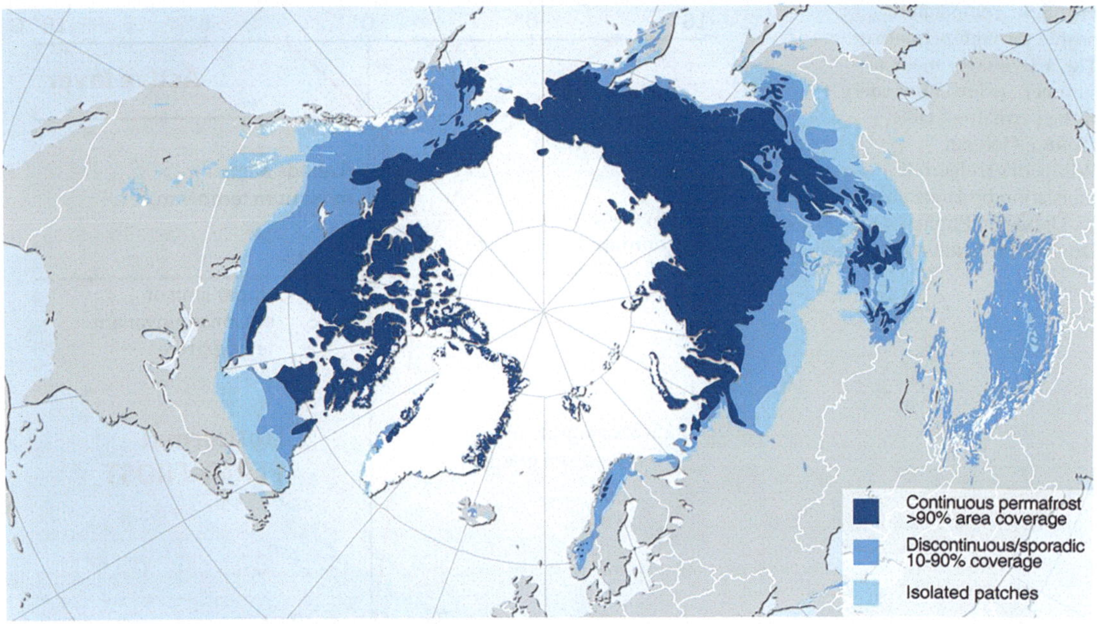

Fig. 5.6 The extent of permafrost in the Northern Hemisphere. Continuous permafrost, discontinuous permafrost, and isolated patches are indicated. (*Source* Hugo Ahlenius/UNEP/GRID-Arendal, https://www.grida.no/resources/5234)

Arctic Archipelago and northern Alaska (Fig. 5.6). In Eurasia, patches of continuous permafrost extend to Mongolia and the Tibetan Plateau, and in North America to the southern parts of Hudson Bay. Also, coastal Greenland, which is not covered by land ice, has primarily continuous permafrost. Discontinuous permafrost that underlies 50–90% of the landscape can be found at the perimeters of large continuous permafrost areas of North America, Asia, and North Europe. Bordering the above-mentioned regions, sporadic permafrost (up to 50% of the landscape) and isolated patches can be found in Iceland and the mountain regions further south, such as the Rocky Mountains, the Alps, the Caucasus, and, in the Southern Hemisphere, the Andes.

A vast majority of global permafrost areas is in the Northern Hemisphere, while the Antarctic continent is mainly covered by the ice sheet. Permafrost does not extend under ice sheets and glaciers, which are effective thermal insulators when thick enough [16]. In Antarctica, permafrost is found in the continental near-shore ice-free areas, in the mountains of the Antarctic Peninsula, and sub-Antarctic islands. An exotic form of perennially frozen ground, the subsea permafrost, can be found underwater on the sea floor, mainly on the continental shelf of the Arctic Ocean in Siberia and Alaska. It is an old remnant from the last Ice Age when the sea level was much lower (about 130 m) than presently, and the regions of the present continental shelves were land, allowing permafrost to develop.

For the permafrost to exist, the annual surface energy balance should not indicate a net heat gain. Consequently, the mean ground surface remains below 0 °C. In the coldest regions near the poles, continuous permafrost exists, but when moving further away from the poles, permafrost becomes discontinuous, sporadic, and finally non-existent. This gradual change is because different thermal conditions exist under the same mean annual air temperatures depending on the surface and ground types. The local characteristics of the surface, including vegetation, strongly impact the characteristics of ground freezing and the extent of permafrost. There is a nearly constant upward geothermal heat flux of

63–67 mW m^{-2} (see [11]) from the non-frozen lithosphere underneath to the permafrost. On top of permafrost, thermal energy is exchanged within the active layer, possibly covered by vegetation and snow, with the atmosphere. These heat fluxes at the permafrost table and its lower boundary, plus soil and ice material properties, define the temperature profile through permafrost and its thickness.

Study Problems

1. Explain that if the annual mean surface (air) temperature increases, the permafrost thins also at the bottom. Note that the zero annual amplitude level does not necessarily change. [Hint: Visualize it as the 'trumpet' temperature curve shifting to the right.]
2. The Finnish Environment Institute has used an empirical formula for the determination of the depth of seasonally freezing ground [17]:

$$h_f = \frac{42\ \mathrm{cm}}{h_s}\mu \cdot b\sqrt{S - 15\frac{h_s}{\mathrm{cm}}\ {}^\circ\mathrm{C}\cdot\mathrm{d}}$$

   where h_s is the snow depth, μ is the soil type coefficient, and b is the field coefficient. Give the physical interpretation of this formula. Compute how the depth of frozen ground varies as the snow thickness increases from 0 cm to 50 cm, and the freezing-day sum is S = 900 °C·d. Compare the cases with (a) A pine forest on moraine ground (μ = 1.15, b = 0.9), and (b) An open field on clay terrain (μ = 0.87, b = 1).
3. The air temperature has an annual cycle with mean temperature $\overline{T}$ and amplitude ΔT. What is the amplitude and time lag of the temperature cycle at 10 m depth in mineral and organic soil?
4. Study the sensitivity of permafrost formation on freezing-degree days and snow accumulation. *Hint:* Use the formula for second year frozen ground thickness.
5. Determine the thermal equilibrium state of taliks.

References

1. Tammiksaar E, Stone IR (2007) Alexander von Middendorff and his expedition to Siberia (1842–1845). Polar Record 43(226):193–216
2. Davis TN (2001) Permafrost: a guide to frozen ground in transition. University of Alaska Press
3. Fox-Kemper B, Hewitt HT, Xiao C, Aðalgeirsdóttir G, Drijfhout SS, Edwards TL, Golledge NR, Hemer M, Kopp RE, Krinner G, Mix A, Notz D, Nowicki S, Nurhati IS, Ruiz L, Sallée J-B, ABA S, Yu Y (2021) Ocean, cryosphere and sea level change. In: Climate change (2021). The physical science basis. Contribution of Working Group I to the Sixth Assessment Report of the Intergovernmental Panel on Climate Change, In: Masson-Delmotte V, Zhai P, Pirani A, Connors SI, Péan C, Berger S, Caud N, Chen Y, Goldfarb L, Gomis MI, Huang M, Leitzell K, Lonnoy E, Matthews JBR, Maycock TK, Waterfield T, Yelekçi O, Yu R, & Zhou B (Eds.). Cambridge University Press
4. Murton JB, Opel T, Toms P, Blinov A, Fuchs M, Wood J, Gärtner A, Merchel S, Rugel G, Savvinov G, Wetterich S (2022) A multi-method dating study of ancient permafrost, Batagay megaslump, East Siberia. Quat Res 105:1–22
5. Seppälä M (1986) The origin of palsas. Geogr Ann A Phys Geogr 68:141
6. Farouki OT (1981) The thermal properties of soils in cold regions. Cold Reg Sci Technol 5(1):67–75
7. Zhang T (2005) Influence of the seasonal snow cover on the ground thermal regime: an overview. Rev Geophysics 43:RG4002
8. Arenson LU, Colgan W, Marshall HP (2015) Physical, thermal, and mechanical properties of snow, ice, and permafrost. In: Shroder JF, Haeberli W, Whiteman C (eds) Snow and ice-related hazards, risks, and disasters. Academic Press, pp 35–75
9. Leppäranta M (2023) Freezing of lakes and the evolution of their ice cover, 2nd edn. Springer
10. Nelson F, Outcalt SI (1983) A frost index number for spatial prediction of groundfrost zones. In: Permafrost fourth international conference proceedings, vol I. National Academy Press, pp 907–911
11. Pollack HN, Hurter SJ, Johnson JR (1993) Heat flow from the earth's interior: analysis of the global data set. Rev Geophys 31:267–280
12. Langer M, Nitzbon J, Groenke B, Assmann L-M, Schneider von Deimling T, Stuenzi SM, Westermann S (2024) The evolution of Arctic permafrost over the last 3 centuries from ensemble simulations with the CryoGridLite permafrost model. Cryosphere 18:363–385
13. Walvoord MA, Kurylyk BL (2016) Hydrologic impacts of thawing permafrost—a review. Vadose Zone J 15:1

14. Opel T, Bertran P, Grosse G, Jones M, Luetscher M, Schirrmeister L, Stadelmaier KH, Veremeeva A (2025) Ancient permafrost and past permafrost in the northern hemisphere. In: Scott E (ed) Encyclopedia of quaternary science, 3rd edn. Elsevier, pp 16–33
15. Smith MW, Riseborough DW (2002) Climate and the limits of permafrost: a zonal analysis. Permafr Periglac Process 13:1–15
16. Tarasov L, Peltier WR (2007) Coevolution of continental ice cover and permafrost extent over the last glacial-interglacial cycle in North America. J Geophys Res Earth 112:1
17. Soveri J, Varjo M (1977) On the formation and occurrence of soil frost in Finland 1955 to 1975 [in Finnish]. Publications of the water research institute 20. National Board of waters, Helsinki

6 Ice Sheets and Glaciers

Landscape photograph showing Vatnajökull glacier, Iceland, in summer. (*Photograph Matti Leppäranta*)

Our interest in ice sheets and glaciers stems from their acting as freshwater storages and potential peril of alpine and coastal habitations as well as from variations of their areal extent and volume in response to climate changes. Also, they serve as an archive of the Earth's climate history. The Antarctic and Greenland ice sheets play a major role in the global climate by regulating the sea

M. Leppäranta, P. Uotila, *Geophysics of the Cryosphere*, Springer Textbooks in Earth Sciences, Geography and Environment, https://doi.org/10.1007/978-3-031-93071-3_6

level and modifying the atmospheric general circulation. The influence of glaciers is limited to the local climatic and hydrological conditions. In this chapter, the basic definitions are first given, and then the structure and classification of glaciers are described, along with a discussion of their geographical distribution and mass balance. Then, the fundamental physical laws governing glacial ice flow, the conservation of mass and momentum, are derived from the basic principles and presented with their applications. Finally, essential aspects of hydrology and water circulation inside and at the perimeters of ice sheets and glaciers are presented, including the calving of icebergs, followed by the introduction to iceberg evolution and drift.

6.1 Formation and Structure of Land Ice Massifs

6.1.1 Classification of Land Ice Massifs

Large land ice formations originate from long-term snowfall. They are found in areas where the snow accumulation in winter does not all melt during summer [1]. Snow that has survived over at least one summer is called *firn*.[1] Next winter, new snow falls on a firn layer, and the whole snowpack can accumulate to a large thickness over the years. Firn is slowly densified and finally transformed into ice through melt-freeze metamorphosis and by the pressure of the overlaying snow. The compression into ice requires a pressure of 0.1–1 MPa or a 50–100 m thick snow layer on top. In this change, the air channels between firn crystals are closed and become sealed gas bubbles with information about the atmospheric constituents at the time. The appearance of this compressed firn resembles ice rather than snow. The pressure deeper in the glacier further compresses the bubbles, making the ice denser.

Large bodies of ice, or ice massifs, are classified into two main categories: *ice sheets* and *glaciers*. Ice sheets are large continental ice masses, also called continental glaciers, with a surface area over 50,000 km^2 and age hundreds of millenniums. Smaller land ice accumulations are counted as glaciers. Ice sheets flow under their weight because of their size, while the terrain slope also forces glacier flow. The tongues of ice sheets and glaciers pushed by gravity to float over the ocean are called *ice shelves*. In the present Earth, there are two ice sheets: the Antarctic (14 million km^2) and Greenland (2.2 million km^2) ice sheets. These are about 4 km thick at their thickest zones and contain about two-thirds of the planet's freshwater resources. They remain mostly unused since they reside far from any major population centres. Sometimes, the East and West Antarctic ice sheets are considered as two individuals.

In the Weichselian Glaciation, northern Europe was covered by the Fennoscandian ice sheet, with the maximum extent about 25,000 years ago. The area was then more than twice the area of the present Greenland ice sheet. At that time, the Laurentide ice sheet covered a large part of North America, and the Patagonian ice sheet covered an elongated zone in the Southern Andes in South America (Fig. 6.1). Relicts of these ice sheets still exist in the Scandinavian Mountain Chain, in Canada, and in the Andes. Elsewhere, even though they disappeared thousands of years ago, signs of their existence are still visible as landscape features, such as eskers, lake districts, and glacial erratics (Fig. 6.2).

Glaciers are found in polar regions at all altitudes but only in mountains in lower latitudes (Fig. 6.3), down to the equator, where glaciers exist in the Andes in Ecuador and Kilimanjaro in Tanzania. At the Arctic Circle in northern Europe, the glaciation level is about 1 km altitude, while tropical glaciers are found above 4–5 km altitude. The largest glacier in the Southern Hemisphere is the Southern Patagonian ice field in Argentina and Chile, covering an area of 13,000 km^2 [2]. In the Northern Hemisphere, the largest glacier is the Severny Island ice cap in Novaya

[1] From Swiss German *firn*, 'last year's'.

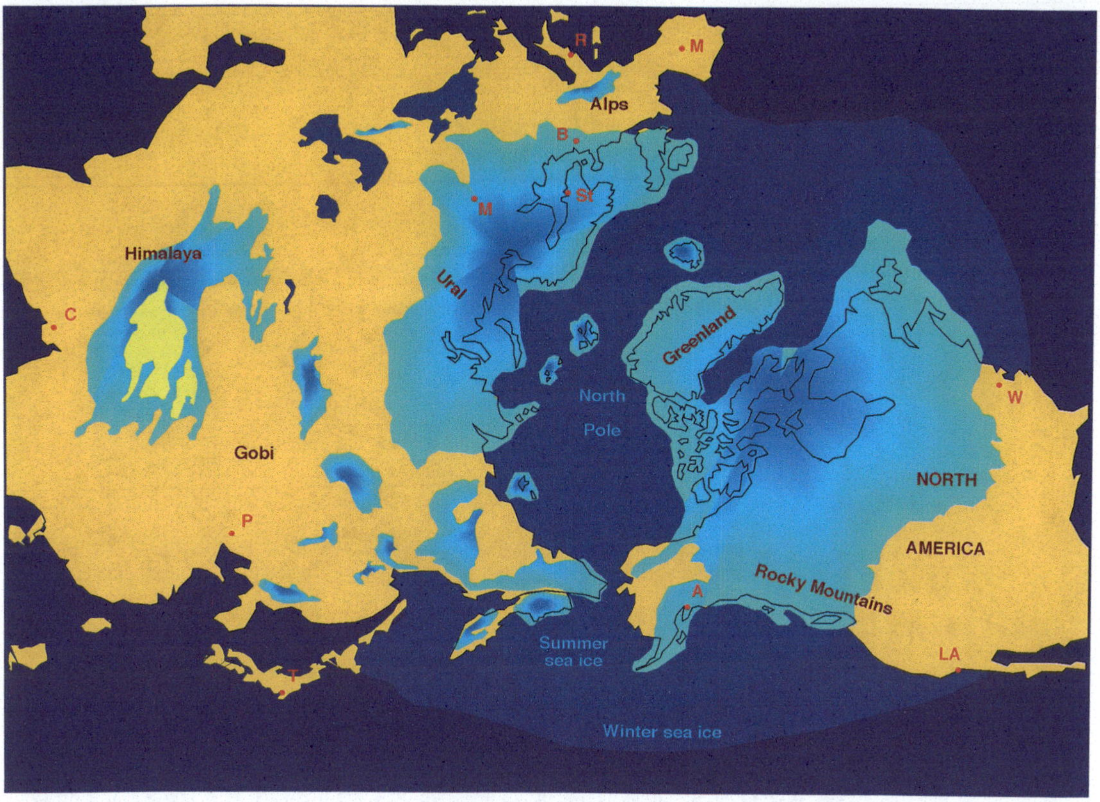

Fig. 6.1 A map on the extent of Northern Hemisphere glaciation (light blue) during the last glacial maximum when 3–4 km thick ice sheets caused a sea level lowering of about 120 m. (*Source* Hannes Grobe/Alfred-Wegener-Institut, copyright CC-BY-2.5)

Zemlya, with an area of 20,500 km^2 (NSIDC[2] website, based on MODIS data), which is 0.9% of the area of the Greenland ice sheet. The next ones are Vatnajökull, Iceland (7900 km^2) and Austfonna, Svalbard (7800 km^2). The thickness of the Vatnajökull glacier is 400–700 m.

Glaciers are classified by their temperature. Beneath the annually active surface layer, the temperature is at the pressure-dependent freezing temperature in *temperate glaciers*, while in *cold glaciers* the temperature is lower than that. The freezing point decreases by about 0.07 °C per pressure of a 100-m layer of solid ice (see Eq. 2.1). Glaciers are also classified as *mountain* and *valley glaciers* for their location, and they can be categorized by their morphology and dynamics (Portrait 6.1).

Example 6.1 In the Scandinavian Mountains, several glaciers are from near sea level to more than 2 km altitude. At Kilpisjärvi, Finland, there are no glaciers even though the highest peak is 1361 m, but nearby at Skibotn, Norway, the Steind alsbreen glacier exists at a lower elevation. The climate is colder, but precipitation is lower in Kilpisjärvi compared with Steind alsbreen, which explains the difference.

Bodies of ice, from small glaciers to massive ice sheets, affect their surroundings in several ways. Naturally, larger amounts of ice have more extensive impacts. The cold snow and ice cool the atmospheric boundary layer above

[2] National Snow and Ice Data Center in Boulder, Colorado.

Fig. 6.2 Large rocks in Turku, Finland, clustered together over 12,000 years ago by an ice flow of the Fennoscandian ice sheet. (*Photograph* Petteri Uotila)

through the surface energy balance. Large air temperature gradients from cold ice massifs towards warm environments and steep slopes induce storm and wind patterns specific to the local climate around ice sheets and glaciers. Like any medium, an ice massif exerts a gravitational pull, and therefore, ice sheets extending to coasts increase the sea level around them. They also press the bedrock, which can result in a depression underneath. When an ice sheet loses mass, the gravitational pull is reduced, and the distribution of glacial meltwater on the ocean surface depends on the change of the Earth's gravity field. The global mean sea level increases if the ice sheet is on land and its meltwater flows to the ocean. Furthermore, the flow of ice erodes the bedrock and carries soil, dirt, and rocks with it (Fig. 6.4). The bedrock underneath an ice sheet rebounds, first faster but slowing down over time and continuing millennia after the decay of large ice sheets. For example, the present rebound of the Fennoscandian ice sheet, 10,000 years after its retrieval, is 1 cm per year at maximum at the location of the past centre of the ice sheet.

6.1.2 Structure of Ice Sheets and Glaciers

The first division of the structure of ice sheets and glaciers shows two parts: the *accumulation zone* and the *ablation zone* (Fig. 6.7). Over an annual cycle, the ice surface gains mass in the accumulation zone, while mass loss occurs in the ablation zone. Typically, the accumulation zone is located in the upper part of the ice massif, and the ablation zone covers the perimeters at a lower elevation where ice decay occurs. The boundary between these zones is the *equilibrium line*, which may change location year by year. Mass

Fig. 6.3 Gorner glacier in the European Alps in summer. Note a group of people walking on the glacier and large pools of meltwater. (*Photograph* Petteri Uotila)

gain is due to precipitation, while ice loss is caused by melting, evaporation/sublimation, and calving of icebergs at the terminus of floating glacier tongues and ice shelves (Fig. 6.5). In a cold climate, sublimation tends to be small. However, sublimation can be the primary mass loss factor in the dry-snow zone where no melting occurs.

The mass budget year consists of the winter accumulation period, when the entire ice sheet or glacier gains mass, and the summer melt period, when the overall mass balance is negative. The annual balance is the sum of winter and summer balances. The flow of ice compensates for the difference between the accumulation and ablation zones. In equilibrium, the compensation is perfect, and the surface topography of the ice sheet or glacier has a fixed annual cycle.

Glaciers can be further divided into subzones based on the processes affecting their mass balance (Fig. 6.6). The coldest zone is the *dry-snow zone*, where there is no melting even in summer. This zone usually exists only in high-polar glaciers. The dry-snow zone ends in the dry-snow line adjoining the *percolation zone*. This zone experiences near-surface melting, and the meltwater percolates into the firn to a certain depth and refreezes, forming ice layers, lenses, and glands. Refreezing releases energy that causes internal heating of the firn. As the summer melting period advances, the percolation deepens, increasing the glacier temperature. The isotherm of the melting point temperature defines the wet-snow line that ends the percolation zone.

In the *wet-snow zone*, all the snow deposited after the previous summer has a temperature at the melting point due to the refreezing by the end of

Portrait 6.1 John Nye (1923–2019) and John W. Glen (1927–2023), right, in front of a glacier [3]. These two highly recognised ice scientists made pioneering contributions to the theory of the flow of ice sheets and glaciers. John W. Glen is known for the glacier flow law named after him. Glen's law resulted from an extensive set of laboratory ice deformation experiments that he carried out as the first scientist in modern glaciology. Furthermore, he defined Glen's relations, or secondary ice flow rates, providing benchmarks against which later results are compared. (*Photograph* John Nye's private collection)

the current summer. Some meltwater can flow deeper than last year's snow layer and refreeze there, but the temperature there is below the melting temperature. This zone ends at the snow line, the boundary between firn and ice at the end of the melt season. Then comes the *superimposed ice zone*, where all ice has merged to a continuous mass due to excess meltwater. The superimposed ice zone ends at the equilibrium line, below which lies the ablation zone.

Most glaciers do not contain all these zones. Especially in temperate and subpolar glaciers, the entire surface can experience melting in summer, and the dry-snow zone, possibly also the percolation zone, is absent. Ice sheets and some cold glaciers, e.g., in Alaska and the Himalayas, contain all the subzones. Within the accumulation zone, ablation patches may appear, e.g., due to wind drift of snow. These patches show up with bare surface, therefore called *blue ice*.

6.1.3 Snow-to-Ice Transition

The density of seasonal snow in a glacier is, at most, about 400 kg m^{-3} influenced by metamorphic processes (see Chap. 4). In the course of years, melt–freeze and compressive metamorphoses further transform the appearance of firn into ice (Fig. 6.7). This transition occurs, by an agreed standard, at the density of 830 kg m^{-3}. At this density, the air channels in the firn have entirely closed into bubbles with a bubble content of about 9% in volume, making a difference between the properties of firn and ice. Thereafter, the ice is packed denser due to the compression of further snow accumulation on top.

In the case of pure compression, an exponential form has been observed for the density of firn vs. depth h [4]:

$$\rho(h) = \rho_{if} - (\rho_{if} - \rho_0)e^{-\lambda h} \qquad (6.1)$$

Fig. 6.4 A glacier edge in the Alps with dirt and rocks. (*Photograph* Petteri Uotila)

where $\rho_{if} = 830$ kg m^{-3}, ρ_0 is the surface density, and λ is the inverse scale depth. This follows from the reasoning that the compression of firn is proportional to the density deficit from ice, $d\rho/dh = \lambda(\rho_{if} - \rho)$. According to observations, λ^{-1}~30–70 m (Fig. 6.8). In the ice beneath the firn layer, the density is

$$\rho(z) = \nu_b \rho_b + (1 - \nu_b)\rho_i \quad (6.2)$$

where ν_b and ρ_b are the porosity and density, respectively, of the bubbles which may contain liquid water. The density of deep glacier ice is about 920 kg m^{-3}.

6.2 Mass Balance

6.2.1 Basic Formula

The local surface mass balance of an ice sheet or a glacier is written as

$$\frac{dm}{dt} = P - E + Y - R \quad (6.3)$$

where $m = m(x, y; t)$ is the mass per unit area, P is the precipitation, E is the evaporation/sublimation, Y is the net transport due to snowdrift, and R is the net runoff including calving of icebergs. Precipitation includes snow, hail and rain, and deposition of hoar is included as negative sublimation. The mass can be taken as the water or ice equivalent in volume per area, e.g., in millimetres of water, per time. Research on the mass balance of glaciers and ice sheets has been done to monitor their evolution due to their freshwater content and role in regulating the mean sea level. Calving of icebergs is a particular field of study in mass balance.

Accumulation occurs mainly on the surface due to precipitation, which is essential to measure, together with its phase and type, in empirical mass balance research. Other factors contributing to the accumulation are wind-blown snow and

Fig. 6.5 Upsala Glacier, from the Southern Patagonian Icefield, calving icebergs into Lake Argentina. Note the calved icebergs drifting out across the lake. Astronaut photograph of Expedition 21 crew, International Space Station, October 25, 2009. (*Source* Earth Science and Remote Sensing Unit, NASA Johnson Space Center https://earthobservatory.nasa.gov/images/41251/upsala-glacier-argentina)

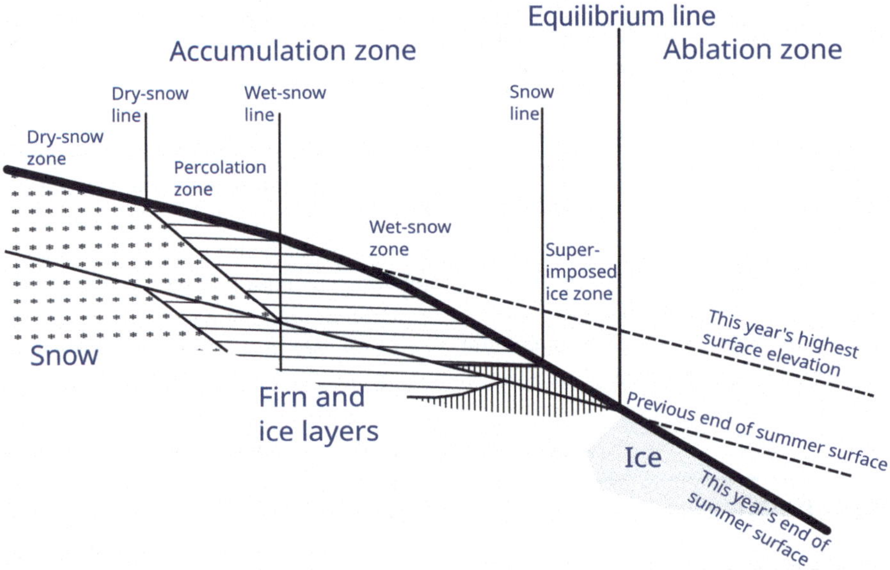

Fig. 6.6 Zones of ice sheets and glaciers illustrated in summer conditions. Equilibrium line separates the accumulation and ablation zones. (Adapted from Cuffey and Paterson [1])

Fig. 6.7 A deep snow pit on a glacier where layers of snow are clearly visible. The site is in a mountain glacier in Queen Maud Land, January 2005. (*Photograph* Matti Leppäranta)

avalanches. Ablation is mainly due to melting in the surface layer and ice–water boundaries in floating sections, where the meltwater flows off the glacier. Typically, meltwater is the dominant loss factor in a mountain glacier, while in Antarctica, most of the loss is due to the calving of icebergs. Furthermore, evaporation and sublimation at the surface, and wind-blown snow leaving the glacier surface contribute to ablation [1].

The net mass balance $b_n = b_n(x, y)$ of a balance year n from the previous year's minimum to this year's minimum is

$$b_n = c_n + a_n \tag{6.4}$$

where c_n is the accumulation and $a_n < 0$ is ablation (Fig. 6.9). Each term in Eq. (6.4) can be presented as a sum of winter and summer balances, $b_n = b_{n,\ w} + b_{n,\ s}$, $b_{n,\ w} > 0$ and $b_{n,\ s} < 0$, etc. In temperate glaciers, mass is gained in winter and lost in summer, so the mass balance cycle has two distinct seasons. Determining the times of annual minima from 1 year to the next is crucial to examining the mass balance change. The days of the minima usually occur in August in the Northern Hemisphere, but the exact calendar date varies between years and glaciers, which complicates the mass balance studies.

In the accumulation zone S_c, $b_n > 0$, while in the ablation zone S_a, $b_n < 0$. By integrating the entire glacier, the net glacial mass balance ΔM is obtained

$$\Delta M = \int_{S_c + S_a} b_n dS \tag{6.5}$$

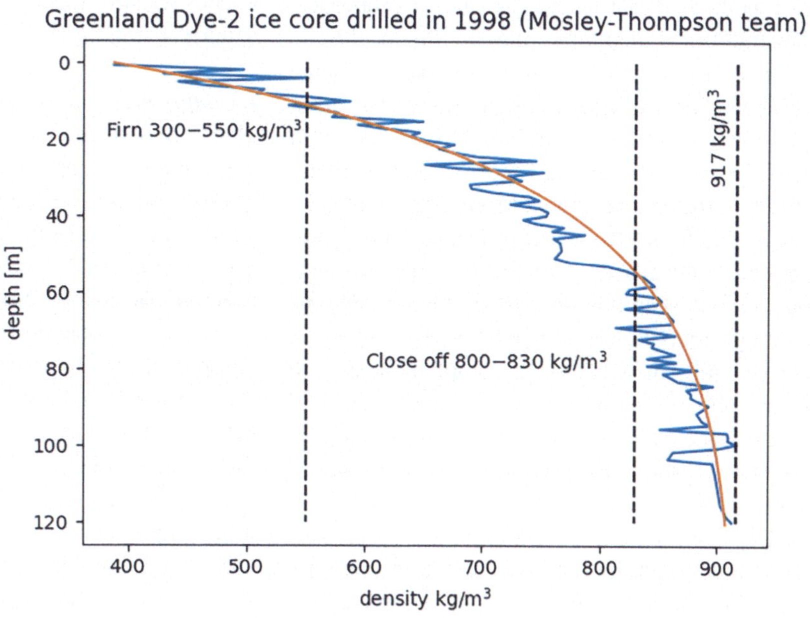

Fig. 6.8 Density profile measured at DYE-2 ice core site in southern Greenland (blue line). The narrow spikes of increased density at the top 40 m are melt layers. At DYE-2, the summer temperatures sometimes get high enough to melt in the surface snow layer. When the liquid meltwater refreezes, it has a higher density than the surrounding firn. The orange line shows a density profile following Eq. (6.1) with the inverse scale depth of 30 m. The density data was obtained from the Byrd Polar and Climate Research Center at Ohio State University

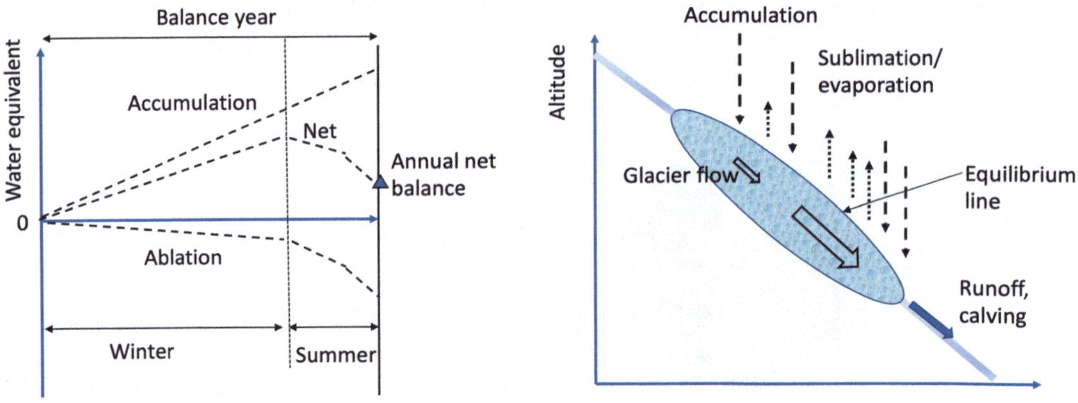

Fig. 6.9 Glacier mass balance diagrams. (**a**) Evolution of annual net mass balance, and (**b**) surface mass balance with altitude

The average net balance $\tilde{b}_n = \Delta M / \int_S dS$ is the net mass balance divided by the glacier area. In a glacier in equilibrium, $\tilde{b}_n = 0$, but the total accumulation and ablation components can be enormous. For ice sheets, they are usually expressed in gigatons per year (about 1 cubic kilometre volume).

6.2.2 Direct In Situ Measurements

Several methods exist to estimate the glacier mass balance and flow [1, 5]. Typically, the most accurate methods directly measure glacier properties. The traditional method to measure the local mass balance and flow is to insert *snow stakes* in the surface layer and measure their surface elevation and location with time. Additionally, a stake's tilt provides information on the vertical velocity shear in the glacier flow. Moreover, snow pits can be dug to determine snow accumulation, temperature, and density profiles from the stratigraphy in the accumulation zone. Such methods, although reliable, are rather laborious and local. Hence, they are sometimes complemented with repeat photogrammetric surveys to validate and calibrate the manual measurements and determine volume changes over multiple mass balance years.

The *hydrological method* is based on the difference between the input and output fluxes (see Eq. 6.3), $\Delta M = P - E + Y - R$. The input is due to precipitation, while the output is due to sublimation/evaporation, meltwater runoff, and ice breakage at the glacial terminus, such as the calving of icebergs. Obtaining the mass balance requires proper knowledge of the three loss components, which can be challenging and crucially dependent on the accuracy of in-situ observations. In mountain glaciers, the mass loss is often dominated by meltwater runoff along one outflow stream, and the discharge can then be easily measured, for example, by dam stations. Sublimation/evaporation can be estimated at least from the routine weather observation network if no glacier site data are available. It is, however, usually a small term.

Example 6.2 The surface area of the Vatnajökull glacier in Iceland is $A = 7900$ km^2 and the annual precipitation is approximately 1000 mm. If evaporation and the accumulation of drifting snow can be assumed to be small, then the scale of the net accumulation is $P_n = 500$ mm. As the net accumulation discharges from one river into the sea, the mean flow rate becomes $P_n A/t = 125$ m^3 s^{-1}, comparable to the discharge of a medium-sized river in northern Europe. The melting season of the glacier is short. If it were 3 months, the mean discharge would be 375 m^3 s^{-1}. In the Icelandic language, a surge of glacial floodwater is called a jökulhaup. One of the most famous cases was the Grímsvötn jökulhaup in 1934; Grímsvötn is a volcano with the centre beneath Vatnajökull.

The peak flow rate was $5 \cdot 10^4$ m^3 s^{-1} (1/4 of the flow rate of the Amazon River), and the extent of the flood was 100 km^2.

6.2.3 Remote Sensing

Remote sensing has become a highly versatile set of tools to map a wide variety of glacier features, such as their surface and bottom topography, structure, and mass and energy balance. In satellite systems, optical and thermal infrared signals originate in the very surface layer, microwave radiometers see into the top layer less than 1 m deep, and at radar wavelengths, the penetration depth may be several metres in dry glaciers. Ground penetrating radar can penetrate through a whole glacier. Although such radars are often employed with antennas posed directly on the surface, they also function well in airborne mode and have been used for mapping the internal structure and bottom topography of glaciers.

Satellite remote sensing has become the primary method of monitoring glacier mass, volume, and topography in recent decades (e.g., [6, 7]). Sequences of aerial photographs or high-resolution satellite images help determine variations in the surficial extent of glaciers. However, there are often difficulties at the lower end of glacial streams where the snout may be buried beneath morainic materials. The surface topography can be determined by airborne or satellite-borne laser profiling that provides the accumulation and ablation. This can also be obtained using satellite-borne radar altimetry, radar interferometry, or high-resolution optical stereograms. Radar interferometric coherence would be an excellent tool for such mapping.

Modern satellite gravimetry, such as GRACE (Gravity Recovery and Climate Experiment), can accurately and directly determine gravitational anomalies due to large glacial mass. Moreover, radar altimetry, such as ICESat (Ice, Cloud, and land Elevation Satellite, NASA), records glacier surface elevation. Glacier Digital Elevation Models (DEMs) are created with optical stereo-imagery or synthetic aperture radar (SAR) interferometry. Repeated measurements can link these three measurement types to ice flow, and mass, or volume, loss, and gain [8].

Contemporary determination of the mass balance often combines observations from multiple sources with suitable methods so that the mass balance is defined comprehensively across the entire glacier. These methods include various types of interpolation or computer models. Such an approach is used mainly for large glaciers and ice sheets. Resulted mass balance estimates typically include uncertainty ranges.

6.3 Flow of Ice Sheets and Glaciers

6.3.1 General Equations

Glacial ice is a viscous medium that flows under the force of gravity. In mountain glaciers, the flow is on inclined surfaces, but ice sheets creep, or deform like a sticky, highly viscous liquid, under their weight. This movement is driven by gravity, and the flow velocity is 10–1000 m per year, or in the order of metres per day. The movement of glaciers can be divided into two types: basal sliding and internal deformation. Basal sliding can be anything between 0% and 100% of the total movement of a glacier.

In geophysical scales, the mechanics of a glacial ice massif can be treated as an incompressible continuum, and the rheology is modelled as a non-linear viscous law (see Sect. 2.2.2). The dynamics of a continuum is governed by the laws of conservation of mass and momentum (Fig. 6.10), i.e., the continuity equation and the equation of motion. These fundamental laws are

$$\frac{\partial u}{\partial x}+\frac{\partial v}{\partial y}+\frac{\partial w}{\partial z}=0, \quad \text{or} \quad \nabla \cdot \boldsymbol{u}=0 \tag{6.6a}$$

$$\frac{\partial \boldsymbol{u}}{\partial t}+\boldsymbol{u} \cdot \nabla \boldsymbol{u}=\frac{1}{\rho} \nabla \cdot \boldsymbol{\sigma}+\boldsymbol{g} \tag{6.6b}$$

where (x, y, z) are the Cartesian coordinates, $\boldsymbol{u} = (u, v, w)$ is the ice velocity, ρ is the ice density, $\boldsymbol{\sigma}$ is the ice stress, and $\boldsymbol{g}$ is the acceleration due to gravity.

Fig. 6.10 Ice flow of the Gorner glacier in the Alps. Streamlines can be seen on the surface characterising the flow structure. This glacier is the second largest one in the Alps, area 53 km^2. (*Photograph* Petteri Uotila)

The continuity equation can be integrated vertically from the glacier bed $h_b(x, y)$ to the surface $h_s(x, y; t)$:

$$w(x, y, h_s; t) - w(x, y, h_b; t) = -\int_{h_b}^{h_s} \left(\frac{\partial u}{\partial x} + \frac{\partial v}{\partial y}\right) dz \approx -H\nabla_H \cdot \tilde{u}_H \quad (6.7)$$

where $H = h_s - h_b$ is the ice thickness, ∇_H is the horizontal gradient operator, and $\tilde{\boldsymbol{u}}_H$ is the vertically averaged horizontal ice velocity. The last step follows the Leibniz integral rule by assuming that the glacier bed and surface slopes are small. Equation (6.7) is a shallow water equation and provides a realistic representation of integral aspect of glacier flow. The vertical velocities at the bed and surface are linked to the changing glacier thickness:

$$\frac{\partial H}{\partial t} = -H\nabla_H \cdot \tilde{\boldsymbol{u}}_H + \Delta b \quad (6.8)$$

where Δb is the net change due to ice accumulation and melting at the surface and bottom.

The velocity of ice is 1–10^3 m per year, and by scaling analysis it is easy to see that the right-hand side dominates the momentum equation (Eq. 6.6b). This represents a quasi-steady system with negligible momentum advection. The driving force of glacier flow is gravity. The resisting forces are internal friction and the bottom friction.

When solving the glacier flow mathematically, no shear stress is assumed at the surface. At the bottom, the slip condition, possibly with a non-zero slide velocity, is taken since the glacier may be sliding on bedrock on a thin film of water because the hydrostatic pressure from the mass of

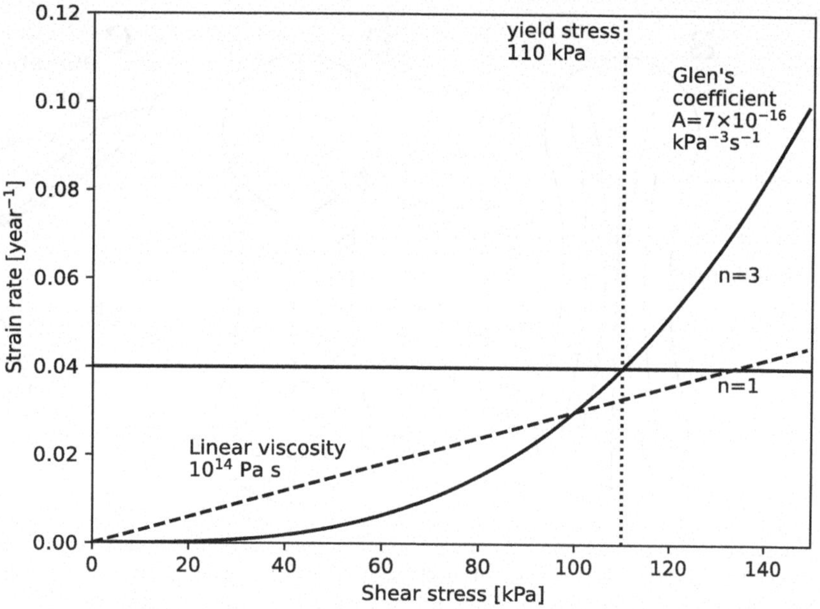

Fig. 6.11 Viscous rheology. In the Glen's glacier flow law, the power $n = 3$ is appropriate. Linear viscous law has $n = 1$, and the limit $n \to \infty$ is plastic rheology

ice above melts the ice at the lower boundary. The glacier flow typically slows down with depth, creating a velocity profile with shear, indicating substantial internal deformation at deep.

6.3.2 Rheology of Glacier Ice

The rheology is a non-linear viscous case. Since incompressibility is assumed, only shear viscosity is relevant as in linear viscous media, but additionally, a form of non-linearity is required (Fig. 6.11). In Glen's viscous flow law [1, 9, 10], a power law form is taken, and the viscosity depends on the deformation invariants in addition to the material properties of glacial ice. In matrix form, Glen's law is expressed as

$$\dot{\boldsymbol{\varepsilon}} = A\|\boldsymbol{\sigma}'\|^{n-1}\boldsymbol{\sigma}' \tag{6.9}$$

$$\boldsymbol{\sigma}' = \boldsymbol{\sigma} - \frac{1}{3}(\sigma_{xx} + \sigma_{yy} + \sigma_{zz})\boldsymbol{I} = \boldsymbol{\sigma} - \frac{1}{3}\sigma_I\boldsymbol{I}$$

where $\dot{\boldsymbol{\varepsilon}}$ is the strain rate, $\boldsymbol{\sigma}'$ is the deviatoric stress, $\|\boldsymbol{\sigma}'\| = \sqrt{\frac{1}{2}\sum\left(\sigma'_{ij}\right)^2}$ is the L_2 norm of the matrix $\boldsymbol{\sigma}'$, A is the inverse viscosity and $\boldsymbol{I}$ is the unit matrix, $I_{ij} = 1(0)$ for $i = j (i \neq j)$. For two dimensions, $^1/_3$ is replaced by ½ in the definition of $\boldsymbol{\sigma}'$. Glen [9] used the exponent $n = 3$ but in general n has varied between 1.5 and 4.2 in different studies. The parameter A depends on n and the ice temperature. For $n = 3$, $A \approx 7 \cdot 10^{-15}\ \mathrm{s}^{-1}\ \mathrm{kPa}^{-3}$ at $T = 0$ °C, and $A \approx 5 \cdot 10^{-17}\ \mathrm{s}^{-1}\ \mathrm{kPa}^{-3}$ at $T = -30$ °C. The deviatoric stress determines the deformation. The full problem requires numerical modelling for the solution, but the following simple shear flow illustrates the physics well.

Example 6.3 Take a glacier in a plane–parallel flow $u = u(z)$ on a fixed mountain slope α. Choose the x-axis along the slope and the z-axis perpendicular with $z = 0$ on the ground surface. Then $g_x = -g \sin\alpha$, $g_z = -g\cos\alpha$, $\partial u/\partial z > 0$, and

$$\dot{\boldsymbol{\varepsilon}} = \begin{pmatrix} 0 & \dfrac{1}{2}\dfrac{\partial u}{\partial z} \\ \dfrac{1}{2}\dfrac{\partial u}{\partial z} & 0 \end{pmatrix}$$

$$= A\|\boldsymbol{\sigma}'\|^{n-1}\begin{pmatrix} (\sigma_{xx} - \sigma_{zz})/2 & \sigma_{xz} \\ \sigma_{zx} & (\sigma_{zz} - \sigma_{xx})/2 \end{pmatrix}$$

where for symmetry $\sigma_{xz} = \sigma_{zx}$ and $\|\boldsymbol{\sigma}'\| = |\sigma_{xz}|$. With $\partial/\partial x \equiv 0$, equation of motion gives $(\nabla \cdot \boldsymbol{\sigma})_x = d\sigma_{xz}/dz$, $(\nabla \cdot \boldsymbol{\sigma})_z = d\sigma_{zz}/dz$. The velocity is solved from the x-component

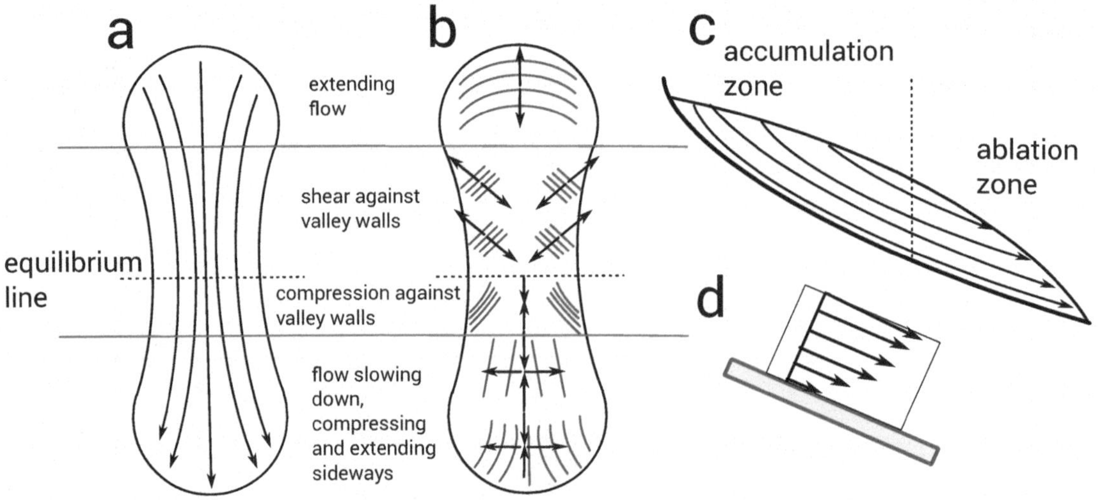

Fig. 6.12 Schematic picture of a valley glacier flow, including distribution and orientation of crevasses. Flow lines are marked as arrows. (**a**) View from above, (**b**) arrows denote horizontal flow gradients, (**c**) sideview, and (**d**) vertical velocity profile and the slip boundary condition. The maximum flow speed occurs at the equilibrium line. (Adapted from https://www.antarcticglaciers.org)

$$\frac{1}{\rho}\frac{d}{dz}\left(\frac{1}{2A}\frac{du}{dz}\right)^{1/n} = g\sin\alpha$$

Using the boundary conditions $u(0) = u_b, \frac{du}{dz}\Big|_{z=H} = 0$ (no shear stress at the surface), the differential equation integrates into

$$u(H) - u(z) = \frac{2A}{n+1}(\rho g\sin\alpha)^n (H-z)^{n+1}$$

Take $\alpha = 10°$ and $n = 3$. In a temperate glacier, $A \approx 7 \cdot 10^{-15}\ s^{-1}\ kPa^{-3}$, and with $H = 200$ m we have $u(H) - u(0) = 1.2$ m day^{-1}. The z-component of the equation of motion is reduced to the hydrostatic pressure law.

Crevasses are fractions in glacier ice due to tension. Their direction and shape indicate how the internal stress acts on ice flow with principal stress-related patterns. Crevasses open in the direction of maximum tension and tensile strain rate, which are related to the largest lateral flow velocity gradients, where the velocity increases in the flow in the accumulation area down the slope. Under such tension, crevasses are oriented perpendicular to the flow direction.

Where the glacier is in contact with surrounding mountains, as in the case of ice flow through a valley, the shear stress against the valley sides generates crevasses along the glacier margin directed with certain angles from the valley sides towards the glacier centre (Fig. 6.12). At the narrowest point of a valley, the idealized glacier flow reaches its maximum velocity, and as the velocity gradients change the sign, crevasses have different shapes and characteristics. This could also occur at the equilibrium line's location if the mass balance allows. After the narrowest valley point, the ice flow becomes divergent, spreading to the sides, inducing tensile stresses in addition to compression along the slope direction as the flow slows down. Such a stress pattern generates crevasses oriented to the valley sides with angles larger than 45°. The ice flow below, towards the terminus, becomes increasingly divergent, with tension parallel to the ice margin and perpendicularly oriented crevasses. Notably, the crevasses move along the ice flow, deform, and rotate. Irregularities and departures from the idealized glacier change crevasses' formation process and characteristics.

Example 6.3 shows that the plane–parallel flow for $n = 3$ has a velocity profile where velocity increases in power 4 with distance from the bottom, while a linear viscous law ($n = 1$) gives a parabolic profile. Thus, with increasing n, the bottom boundary layer becomes shallower. At the limit $n \rightarrow \infty$, a plastic flow occurs with a sharp slip at the boundary.

Plastic rheology can be used as a convenient approach in analytic analyses. Then, a glacier stands rigid when the shear stress is below the yield stress but deforms at the yield stress. On a mountain slope, the shear stress at the bottom is $\rho g H \sin \alpha$, which reaches the yield stress σ_S~100 kPa when $H = \frac{\sigma_S}{\rho g \sin \alpha}$. For $\alpha = 5°$, $H = 100$ m. Consider the profile $H = H(x)$ of an ice sheet on an even terrain with maximum elevation at $x = 0$. The shear stress at the bottom and the boundary condition are

$$\sigma_{xz} = \rho g H \frac{dH}{dx}; \quad H(L) = 0 \tag{6.10}$$

where L is the ice sheet terminus. The equilibrium solution for $\sigma_{xz} \equiv \sigma_s$ reads $H^2 = 2\frac{\sigma_S}{\rho g}(L - x)$, i.e., the ice sheet has a parabolic profile. For example, when $L = 1000$ km, we have $H(0) = 4.5$ km. Thus, when the size of a continent or island is given, the shear strength of glacier ice determines the size of the ice sheet or glacier that can be built up.

6.3.3 Shallow Ice and Shelf Approximations

Take the steady-state equation of motion (Eq. 6.6b) with the advection term neglected. Two commonly used approaches exist for the horizontal components of glacier flow: shallow ice (SIA) and shallow shelf (SSA) approximations, of which SIA is more common and suitable for describing the ice flow of grounded ice sheets. In the vertical component of Eq. (6.6b), the vertical normal stress dominates, and this component can be integrated to

$$0 = \int_z^{h_s} \left(\frac{\partial \sigma_{zz}}{\partial \zeta} - \rho g \right) d\zeta = \sigma_{zz}(h_s) - \sigma_{zz}(z) - (h_s - z)\rho g \tag{6.11}$$

considering the gauge pressure, $\sigma_{zz}(h_s) = 0$, and $\sigma_{zz} = -(h_s - z)\rho g$. This stress represents the hydrostatic pressure in the system.

In the SIA model, the horizontal gradient of the shear stress is assumed to be small compared to the gradients of the normal stress and the vertical gradient of shear stress [11]. The x-component of the momentum equation is

$$\frac{\partial \sigma_{xz}}{\partial z} = \rho g \frac{\partial h_s}{\partial x} \tag{6.12}$$

The y-component follows similarly. The vertical gradients of shear stresses are proportional to the glacier surface slopes. Since shear stresses are positively related to velocity gradients, in simple shear $\sigma_{xz} \propto (du/dz)^{1/n}$ so that the vertical shear of horizontal velocity is higher, whereas the glacier surface slope is steeper.

The shallow shelf approximation (SSA) could complement the SIA approximation and works best for floating ice shelves for which the vertical shear is small and longitudinal stresses dominate. The SSA approximation is for the x-component

$$\frac{\partial 2\sigma_{xx} + \sigma_{yy}}{\partial x} + \frac{\partial \sigma_{xy}}{\partial y} = \rho g \frac{\partial h_s}{\partial x} \tag{6.13}$$

and correspondingly for the y-component follows. The velocity vectors are not directed parallel to the surface but slightly downward in the accumulation zone and upward in the ablation zone (Fig. 6.12).

Gravity pulls a glacier down in two ways: because of the sloping bedrock underneath and because of the surface slope of the glacier that introduces horizontal pressure gradients inside the glacier. As a result, the surface slope determines the driving stress, and ice flow tends to be directed towards the maximum slope. On the other hand, maps of glacier surface contours can be used to determine glacier flow lines. The driving stress can be calculated from slope and glacier thickness H,

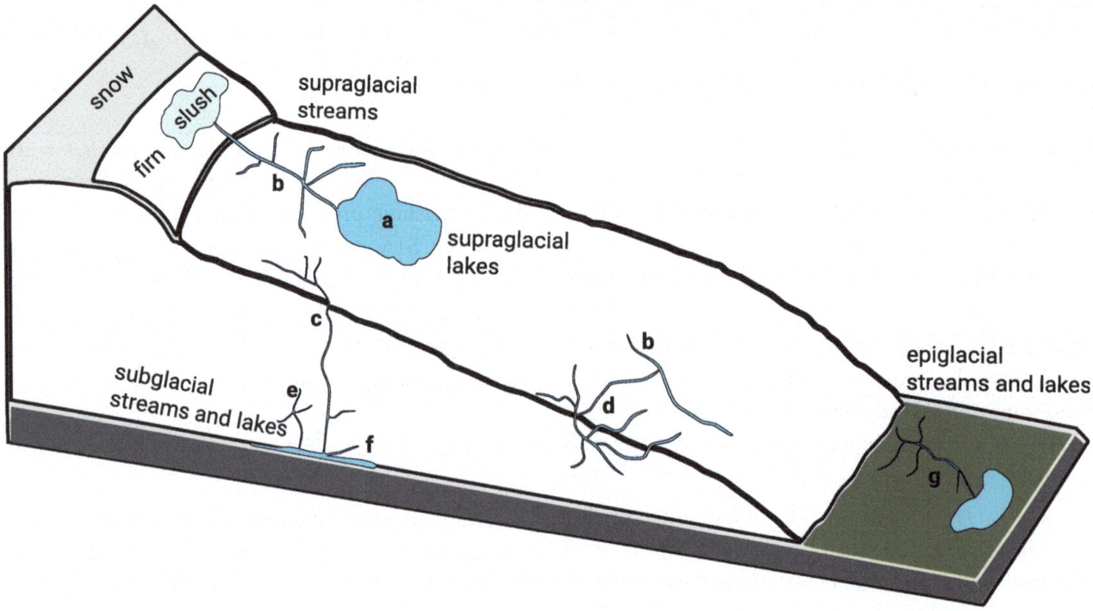

Fig. 6.13 Glacier water system components: (a) supraglacial lake, (b) surface streams, (c) moulins draining into subglacial tunnels, (d) crevasses partially filled with water, (e) fractures filled water, (f) subglacial hydrological system, and (g) runoff from the glacier. (Adapted from Fig. 1 of the Wikipedia article on Glacial Stream © CC BY-SA 4.0)

$$\sigma_b = f'\rho g H \sin\alpha \approx f'\rho g H \alpha \tag{6.14}$$

as the slope angle α is typically small. The basal stress ranges from 50 to 150 kPa, which provides a yield stress for the assumed plastic ice of 100 kPa [12]. As the basal stress is around the yield stress, the approximation that $H\alpha$ is close to constant is valid. This means that the glacier is relatively thin where the slope is steep, but the glacier is thick where the slope is small.

6.4 Glacial Hydrology

6.4.1 General

Water storage and flow in glaciers are analysed by glacier hydrology. It is an essential discipline for many reasons, including the formation of rivers and lakes fed by glacier meltwater, the use of glacial streams for hydropower production and a source of drinking water, and potential hazards if the outflow suddenly becomes very large, especially in densely populated regions, such as India and Pakistan. Meltwater-related features, such as proglacial lakes, channels and moulins, form a hydrological network where water flows on and inside the glacier (Fig. 6.13).

Glaciers contain liquid water with a volume V_w. The water budget equation reads

$$\frac{dV_w}{dt} = S_w - Q_w \tag{6.15}$$

where S_w is the inflow and Q_w is the outflow. The inflow consists of meltwater and liquid precipitation, and the outflow is mainly due to runoff but also, to a smaller extent, due to evaporation. The outflow depends on the glacier structure and is assumed to be proportional to V/τ_R, where τ_R is the residence time of water in the glacier, which also varies with time.

Fig. 6.14 A supraglacial lake at Aboa research station, Queen Maud Land [13]. The bare surface is persisted by turbulence in the lee side of the dominant wind over the Basen nunataq. (*Photograph* Matti Leppäranta)

6.4.2 Proglacial Lakes

Proglacial lakes exist on, under and at ice sheets and glaciers. *Supraglacial lakes* form on the surface layer of blue ice spots due to melting by solar radiation (Fig. 6.14). The bare, blue ice has a low albedo to absorb much solar radiation and is impermeable in the solid phase. A supraglacial lake can undergo a stable annual cycle of refreezing in winter, or it can grow year-by-year and, due to the water pressure, finally crack the ice to force the water to flow into the glacier through the cracks and vertical channels in a hydraulic form called moulins. Their location needs to be suitably flat with a gentle slope. Seasonal supraglacial lakes are shallow, with a depth of about 1 m. Their surface is typically open in summer in a warm climate but can have a thin ice cover in a cold environment. Supraglacial lakes are found in Greenland and Antarctica and on mountain glaciers.

In places, water can get trapped and accumulate in depressions along the glacier edges. Lakes formed this way are *epiglacial*. Their size varies greatly, from small ponds to large lakes with depths of a few hundred metres. Epiglacial lakes are the most common type of proglacial lakes in Antarctica. At the end of the Last Ice Age, vast epiglacial lakes existed, Lake Agassiz in North America and the Baltic Ice Lake in North Europe. Meltwaters from the former could shut down the Atlantic Meridional Overturning Circulation after being released to the North Atlantic. As a comparison, a contemporary large epiglacial lake, Lake Untersee, is seen in Antarctica, with a maximum depth of 169 m, length of 6.5 km, and width of 2.5 km.

Deep beneath the Antarctic ice sheet, an ancient and largely unknown network of subglacial lakes and rivers is present. This hydrological network is also found under the Greenland ice sheet and in mountain glaciers. The largest subglacial lake is Lake Vostok, with a characteristic lateral size of 100 km and depth of 100 m. Located on the bedrock, geothermal heat flux Q_g affects their depth, while the insulating properties of the overlying ice limit the upward heat loss from the lake. The ice sheet exerts immense pressure p on the subglacial water body, reducing the freezing point temperature $T_f = T_f(p)$. In a steady state, geothermal heat is conducted up from the subglacial lake through the ice to the atmosphere. For the ice sheet thickness of 4 km, $T_f \approx -2.5$ °C (Eq. 2.1), and at the surface temperature −50 °C, the steady-state heat conduction through the ice would be around 25 mW m^{-2}, which matches the information on the geothermal heat flux.

Subglacial hydrological networks are essential in regulating glacier flow through frictional effects between the ice and bedrock. Water lubricates the interface, reduces the friction, and speeds up the flow. At extreme, hydrological processes can contribute to glacier flow instabilities and surges.

6.5 Icebergs

Icebergs are calved from the terminus of ice shelves and floating glacier tongues. Locally, cold icebergs cool the surrounding air and form fog. Their calving contributes to glacier mass loss and, therefore, forms a component of the mass balance. Melting icebergs in the ocean add fresh water to the upper ocean and, hence, influence the local stratification and stability of the water column.

The size of icebergs varies depending on their origin and evolution. Large icebergs are calved off ice sheets, while small glaciers produce small icebergs. Typically, the diameter of an iceberg is several kilometres, and the thickness is 200–300 m [14]. According to the Archimedes principle and iceberg shape, a 30–50-m top layer of such iceberg forms the freeboard above the sea surface. Ice islands are tabular-shaped icebergs up to 60 m thick and 30 km in diameter, originating from the northern glaciers of Ellesmere Island in the Arctic Ocean. A famous one was T-3 or Fletcher's ice island, which was used as a research base in 1952–1978 in the Arctic Ocean.

Icebergs can be massive. A cylindrical iceberg with a lateral diameter of 2500 m and thickness of 300 m has a mass of about five gigatons ($5 \cdot 10^{12}$ kg). Due to their massive size, irregular shape, and the fact that most of their volume is below the sea surface, icebergs pose a serious risk for sea-going vessels.

Icebergs are classified into three main categories by their shape. An iceberg is tabular if the length/height ratio is more than five; otherwise, it is irregular or rounded. In these three main categories, some subcategories give the shape more precisely, such as dome, pyramid or jagged. Icebergs are also classified for their size: very large, large, medium, and small. An iceberg is bigger than a merchant ship (or more than 5 m above the sea level), a bergy bit has the size of a small house (1–5 m above the sea level), and a growler has the size of a grand piano (less than 1 m above the sea level). In particular, growlers pose a risk for seafarers as they are difficult to observe from a ship.

The most enormous icebergs calve off from the Antarctic ice sheet and are tabular in shape. They can be up to 500 m thick and 100 km in their lateral scale and drift from the polar seas to 40–45° latitude. The largest by area recorded was iceberg B-15 that calved from the Ross Ice Shelf in March 2000 (Fig. 6.15). Then it measured 295 × 37 km^2 but broke into several pieces in 2000–2003. Subsequently, the largest piece, B-15A, drifted slowly west in the coastal East Wind Drift and broke into several pieces in 2005 off the East Antarctic coast. In 2018, four pieces of B-15A remained, the northmost one B-15Z, reaching the latitude 54°S close to South Georgia in May 2018 after 18 years of travel.

The evolution of B-15 is quite typical for an Antarctic iceberg. After calving, they tend to ground on the shallow continental shelf, just off the ice sheet margin. After some erosion, scouring and perhaps breakage, they start drifting west in the East Wind Drift. When the coastal current connects to the eastward flowing Antarctic

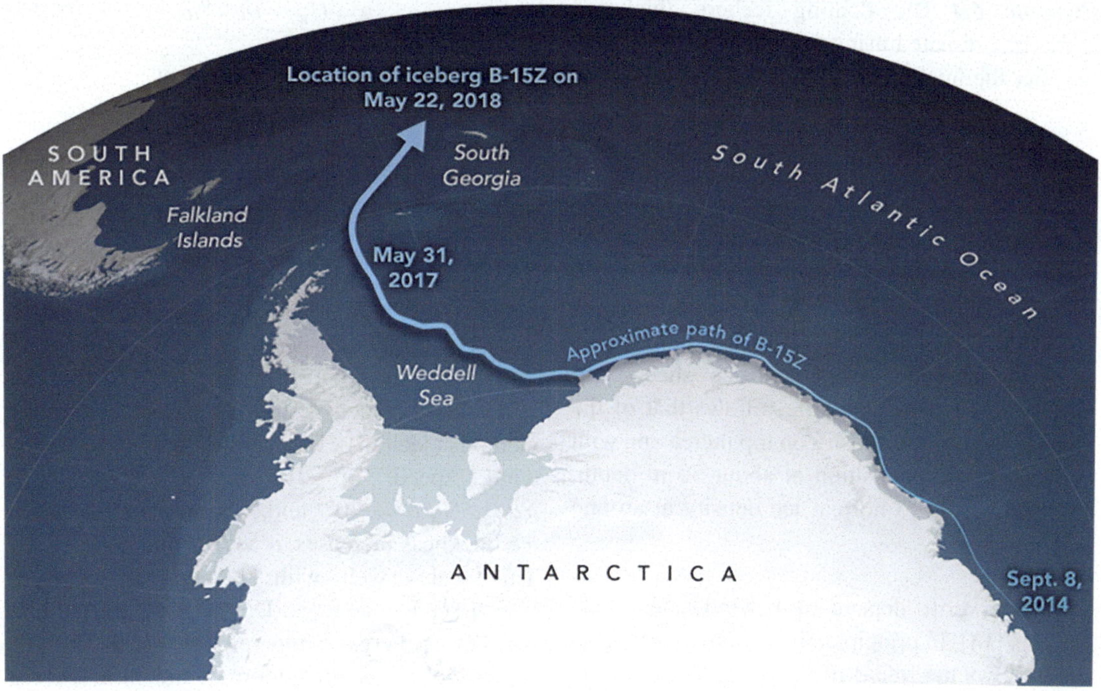

Fig. 6.15 Drift of B-15 iceberg in 2000–2018. Calved off the Ross Ice Shelf in March 2000, it was the world's largest recorded iceberg, with a size of 295 × 37 km^2. (*Source* NASA)

circumpolar current (ACC) with an offshore northward flow, such as in the Weddell Gyre, the Ross Sea Gyre and the Kerguelen Plateau, icebergs leave the coast and move north and are captured by the ACC. In the ACC, they tend to move northeast, forced by the ACC and Coriolis effect promoted by their great thickness.

In the Northern Hemisphere, icebergs are most frequent in the west North Atlantic, in the Labrador Sea, and in Baffin Bay. This is a result of their origin in the Greenland ice sheet. Some ice islands originate from glaciers terminating the Arctic Ocean, but they usually drift to the North Atlantic via Fram Strait to the cold East Greenland Current. They join the icebergs that have calved from the East Greenland ice streams. After the East Greenland Current curves around Cape Farewell in the southern tip of Greenland and becomes the West Greenland Current, iceberg sources of the mighty West Greenland glaciers release even more icebergs to the coastal current. Eventually, icebergs turn south in the Baffin Bay and continue their southward drift past Newfoundland before finally melting in the North Atlantic.

Due to their large volume, icebergs can reach low latitudes before melting. If an iceberg is captured in a cold core ocean eddy moving equatorward, it can travel particularly far towards the equator before melting. Antarctic icebergs have been observed at around 50°S in the waters of New Zealand, originating from the Ross Ice Shelf and South Georgia in the northern Weddell Sea. In November 2006, icebergs were sighted off the coasts of Otago and Canterbury, New Zealand at 46°S. In the North Atlantic, icebergs can travel as far south as 40°N, such as the Grand Banks, east of Newfoundland. S/S *Titanic* sank near there at 41.73°N 49.95°W on April 15, 1912. The furthest recorded sighting occurred near Bermuda at 32°N in 1926. During their voyage, they pose a risk to ships and fixed structures, such as oil and gas platforms.

Example 6.4 The floating iceberg thickness H can be estimated using the Archimedes principle after the freeboard H_f has been observed

$$H = \frac{\rho_w}{\rho_w - \rho_i} H_f$$

Given the density of seawater, $\rho_w = 1025$ kg m^{-3}, and the density of ice, $\rho_i \approx 900$–920 kg m^{-3}, results in $H = 8.2H_f - 9.8H_f$. If an iceberg with a freeboard of 30 m is observed, it can be expected to extend down to 220–260 m depth. Because icebergs are calved off from ice sheets and glaciers, their density profile follows that of the glaciers, with snow density on top increasing with depth to firn/ice transition at about 50 m depth, and then finally to normal ice density at around 100 m.

Iceberg drift depends on wind and ocean currents [14]. In principle, their momentum equation follows the free drift of a very thick sea ice floe (see Sect. 8.5.2), but for icebergs the ice–ocean drag force is a highly complex skin drag and form drag combination, which acts as the main driving force. The equation of motion is expressed, using complex vectors, as

$$\rho H\left(\frac{\partial u}{\partial t} + ifu\right) = \tau_a + \tau_w - \rho Hg\nabla p \tag{6.16}$$

where $u = u_x + iu_y, i = \sqrt{-1}$ is the iceberg velocity, f is the Coriolis parameter, τ_a and τ_w are the lateral stresses of wind and water on the iceberg, and p is the water pressure. The approach is the most appropriate for tabular icebergs, while the drift of irregularly shaped icebergs is more difficult to predict due to the form drag.

The turbulent wind and water stresses are, in general, quadratic in air–ice and water–ice velocity differences, respectively. The water velocity changes with depth, and we need to use a convenient, representative velocity, say, U_w for that. The same problem exists with the pressure gradient, and using the geostrophic flow formula, we can write $\rho Hg\nabla p = -i\rho Hf U'_w$. Assuming further that $U'_w \approx U_w$ (as is done in sea ice dynamics), steady-state case becomes

$$u = U_w + \alpha e^{-i\theta} u_a \tag{6.17}$$

$$\begin{aligned} if\rho H(u - U_w) = {} & \rho_a C_a |u_a| u_a \\ & - \rho_w C_w e^{i\theta_w} |u - U_w|(u - U_w) \end{aligned}$$

where α and θ are the wind factor and deviation angle in wind-driven iceberg drift C_a and C_w are the drag coefficients of air and water, respectively, u_a is surface wind velocity, and θ_w is the turning angle between the ice–water stress and the representative wind speed.

Solving Eq. (6.17) leads to a fourth-order polynomial, which is easy to solve by iterative methods (see [13]). For ice thickness of 10 m, the wind speed of 10 m s^{-1}, we have $\alpha \approx 1.5\%$, $\theta = 70°$, and, asymptotically, as the ice thickness increases, $\alpha \propto H^{-1}$ and $\tan\theta \propto H^2$. This scales well with the typical value of $\alpha = 0.1\%$ for icebergs. Due to the higher thickness of icebergs compared with sea ice, the Coriolis effect is much more dominant in iceberg drift. The iceberg wind factor is lower, and the turning angle is higher. The drag coefficients of icebergs are expected to vary largely between different icebergs.

During their drift to a warm environment, icebergs gradually disintegrate by breakages and melting. Several physical mechanisms contribute to the iceberg decay process. The strong Coriolis effect gives a large northward component to iceberg drift in the Antarctic Circumpolar Current, and in the Arctic, the Labrador Current transports Greenland icebergs south. The freeboard is exposed to warm air, winds, and solar radiation, resulting in surface layer melting, and meltwater percolation and refreezing. Moreover, the oceanic surface wave action scours icebergs at the water level, and the swell may bend and break icebergs. Underwater melting by warm seawater can vary with depth, depending on the ocean stratification, and can cause the capsizing of icebergs. This is common in strongly stratified fjords. Fjord dimensions also affect when icebergs leave the fjord as wide fjords tend to have a gyre circulation that makes it easier for icebergs to depart compared to narrow fjords with an estuarine type of circulation.

A rough rule-of-thumb formula for iceberg melting is given by [14]: $\Delta r = K \cdot T_w$ where Δr is the loss of ice in the walls and bottom of an iceberg, $K \sim 0.12$ m (°C day)$^{-1}$ is a constant, and T_w is the mean water temperature in °C over the iceberg draft. The formula highlights the dominant effect of oceanic melting. For example, if the seawater temperature is 4 °C, $\Delta r = 0.48$ m day^{-1}, and one can expect a loss of 175 m in a year.

Iceberg meltwater fluxes substantially affect ocean stratification, circulation, and climate. For example, the freshwater flux from melting Antarctic icebergs has been estimated to be approximately 0.05 Sv (1 Sv = 10^6 m^3 s^{-1}), which is comparable to the discharge of the World's second largest river Ganges, 0.044 Sv [15]. With this respect, in addition to the impact of icebergs on ocean stratification, circulation and sea level, they represent a significant component of the freshwater budget of the ocean.

Study Problems

1. Explain which zones (or facies) exist in a glacier and their significance. Your answer should be such that it would be a good answer in an exam!
2. Given the current annual mean precipitation (120–150 mm) and mean thickness of the ice sheet (2 km), determine the Antarctic ice sheet's recovery time scale.
3. Derive the exact form of the vertically integrated continuity equation (Eq. 6.7) by using the Leibniz integration rule for the integration limits.
4. Show that the strain rate and stress deviator norms satisfy $\|\dot{\boldsymbol{\varepsilon}}\| = A\|\boldsymbol{\sigma}'\|^n$ in the Glen's rheology $\dot{\boldsymbol{\varepsilon}} = A\|\boldsymbol{\sigma}'\|^{n-1}\boldsymbol{\sigma}'$.
5. An iceberg vertical profile has snow (400 kg m^{-3}) at the surface, then firn (800 kg m^{-3} at 50 m) and ice (917 kg m^{-3}) in the iceberg core just below the water line. Assuming linear density increases, determine the freeboard of the iceberg.

References

1. Cuffey K, Paterson W (2010) The physics of glaciers, 5th edn. Butterworth Heinemann/Elsevier, Oxford
2. Rignot E, Rivera A, Casassa G (2003) Contribution of the Patagonia Icefields of South America to sea level rise. Science 302(5644):434–437
3. Berry M (2020) John Frederick Nye. 26 February 1923 – 8 January 2019. Biogr Memoirs Fellow R Soc 69:425–441
4. Schytt V (1958) The inner structure of the ice shelf at Maudheim as shown by core drilling. In: Norwegian–British–Swedish Antarctic expedition, 1949–52, scientific results 4, glaciology 2, pp 115–151. Norwegian Polar Institute, Tromsø
5. Holmlund P, Jansson P (1999) The tarfala mass balance programme. Geogr Ann Ser A Phys Geogr 81(4): 621–631
6. Pellikka P, Rees G (eds) (2010) Remote sensing of glaciers—techniques for topographic, spatial and thematic mapping of glaciers. CRC Press, Boca Raton
7. Rees WG (2006) Remote sensing of snow and ice. CRC Press, Boca Raton
8. Berthier E, Floricioiu D, Gardner AS, Gourmelen N, Jakob L, Paul F, Treichler D, Wouters B, Belart JMC, Dehecq A, Dussaillant I, Hugonnet R, Kääb AM, Krieger L, Pálsson F, Zemp M (2023) Measuring glacier mass changes from space—a review. Rep Prog Phys 86(3):036801
9. Glen JW (1958) The flow law of ice. A discussion of the assumptions made in glacier theory, their experimental foundations and consequences. IASH 47:171–183
10. Nye J (1959) The motion of ice sheets and glaciers. J Glaciol 3:493–507
11. Zhang Z, Moore JC (2015) Glaciers and sea level rise. In: Zhang Z, Moore JC (eds) Mathematical and physical fundamentals of climate change, Chapter 13. Elsevier, Amsterdam, pp 441–455
12. Li H, Ng F, Li Z, Qin D, Cheng G (2012) An extended "perfect-plasticity" method for estimating ice thickness along the flow line of mountain glaciers. J Geophys Res Earth 117:F01020
13. Leppäranta M (2011) The drift of sea ice, 2nd edn. Springer, Berlin
14. Wadhams P (2000) Ice in the ocean. Gordon & Breach Science, London
15. Depoorter MA, Bamber JL, Griggs JA, Lenaerts JTM, Ligtenberg SRM, van den Broeke MR, Moholdt G (2013) Calving fluxes and basal melt rates of Antarctic ice shelves. Nature 502:89–92

7 Lake Ice and River Ice

A 7-km-long ice road across Lake Pielinen in Finland. It is annually opened when ice thickness is 40 cm. Traffic rules: maximum weight is 3 tons, maximum speed is 50 km h^{-1}, minimum distance between cars is 50 m, and no stop or overtaking. (*Photograph Ms. Maritta Räsänen, Pohjois-Savo Centre for Economic Development, Transport and the Environment, Kuopio. Printed with permission*)

In this chapter, the physics of ice cover in lakes and rivers is presented. Lake ice questions are largely local due to the limited size of the basins. River ice differs from lake ice due to river

M. Leppäranta, P. Uotila, *Geophysics of the Cryosphere*, Springer Textbooks in Earth Sciences, Geography and Environment, https://doi.org/10.1007/978-3-031-93071-3_7

hydraulics which produces large volumes of frazil ice and drives ice jams in ice decay period. Most lakes have fresh water, salinity less than 0.5‰ (per mille); it can be a significant factor in the melting period but otherwise it has a negligible influence on winter processes and ice properties. Saline lakes can be studied with the methods introduced for sea ice in the Sea Ice chapter. Here, the first section introduces freezing lakes and rivers, followed by ice formation and structure and properties of ice. Sections 7.3 and 7.4 present thermodynamics and mechanics of lake ice, and the last section focuses on river ice questions.

7.1 Introduction

7.1.1 Freezing Lakes and Rivers

Frozen inland waters have been present in people's winter environment in cold regions [1–3]. Solid ice covers formed a major part in winter road networks in North Europe from the Stone Age until the nineteenth century. It was easy also to cross on snow-covered land between water courses—a major limitation in boat traffic. Frozen freshwater basins provided an easy way to cut and store ice blocks in winter for cooling food in summer. This ice business ended at mid-twentieth century when refrigerators took finally over. Ice-covered waters have been important bases for local recreation activities, winter festivals, and tourism. Safety questions are then a major issue, as well as the quality of ice and eventual snow cover on ice.

The low-latitude edge of potentially freezing inland waters can be taken as the January 0 °C isotherm in the Northern Hemisphere (Figs. 7.1 and 7.2). The ice zone covers most of Eurasia and North America down to 40°N – 50°N at sea level. In the Central Asian mountains, the boundary is further south, while in the Western Europe, it is more to the north due to the warm North Atlantic Current. Apart from the Antarctic continent, in

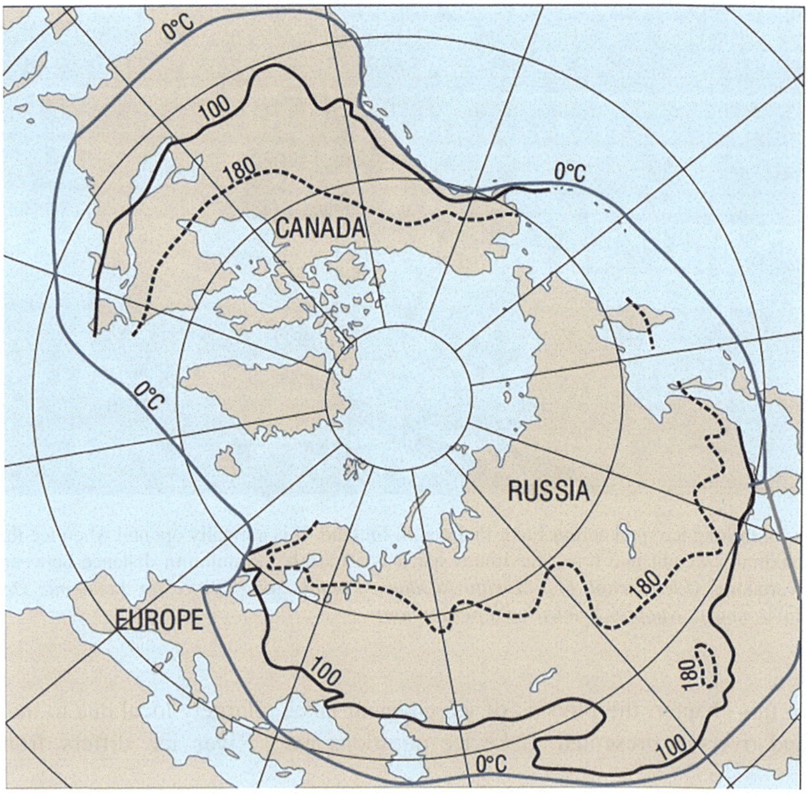

Fig. 7.1 The zone of seasonally freezing inland waters in the Northern Hemisphere. The contours 100 and 180 refer to the mean length of ice season (days), and 0 °C to January mean air temperature [2, 4]

a b

Fig. 7.2 Examples of freezing lakes and rivers. (**a**) An ice-covered lake in the Beiluhe Basin, Qinghai–Tibet Plateau at 4600 m altitude, (**b**) a river in Urho Kekkonen National Park, Northern Finland. The lake has a clear ice cover, with scattered gas bubbles in the structure. (*Photographs* **a** Zhijun Li, **b** Urpo Häyrinen, Finnish Forest Museum Lusto. Printed by permission)

the Southern Hemisphere, only some Southern Ocean islands and mountains in South America and New Zealand contain freezing inland waters. Perennial lake-ice is very rare. In Nunavut and Alaska, remnants of ice survive over summer in some lakes have, and in the Dry Valleys of Antarctica, permanent bare perennial lake ice can be several metres thick. All lakes do not freeze in the cold climate zone: a deep lake may have thermal memory longer than the winter, geothermal heating can compensate for surface heat losses, and salinity depresses the freezing point.

In limnology, lakes smaller than 50,000 m^2 are defined as ponds. A characteristic 'medium lake' has the areal scale of 10 km^2 and depth scale of 10 m. The sizes of small and large lakes can be taken as 1 and 100 km^2, respectively, while deep lakes are deeper than 100 m. Rivers are characterized by permanent, gravity-driven flow, and shallow depth. In quiet parts, such as semi-isolated bays, ice season can be as in lakes, but otherwise the ice conditions are governed by river geometry and turbulent flow. The largest of freezing rivers is Yenisei in Siberia with discharge of about 20,000 $m^3\ s^{-1}$, length 5500 km, and drainage basin 2.45 million km^2. Typical rivers in the Baltic Sea basin have 100–500 $m^3\ s^{-1}$ discharge, 100–500 km length, and 10,000–50,000 km^2 drainage basin.

7.1.2 Cooling and Freezing of Lake and River Water

The density of lake and river water or the equation of state is written in general form as $\rho_w = \rho_w(T, S, p)$, where T is the temperature, S is the salinity, and p is the pressure (Annex A1; [5]). Salinity is usually ignored in fresh water ($S \leq 0.5‰$), and the influence of pressure is significant at $p > 1$ MPa (water depth > 100 m). Freezing point (T_f) and the temperature of maximum density (T_m) are the critical temperatures during cooling, equal to, respectively, 0 and 3.98 °C for pure water and −0.03 and 3.87 °C for $S = 0.5‰$ (Annex A1).

In fall, surface waters are vertically mixed during cooling until around $T = T_m$. Then, buoyancy tends to stabilize the stratification while turbulence tends to continue mixing. The period $T_f < T < T_m$ was called *pre-winter* in Kirillin et al.

[6] since the water surface can potentially freeze fast on a thin, cold surface layer. In shallow lakes and rivers, the mixed layer may keep to the bottom until freezing. Heat storage is proportional to the mixed layer depth, which is the principal water body characteristic to influence the timing of freezing. The lateral lake size scales the wind fetch to drive surface waves and mixing, and therefore large lakes freeze later than small ones.

A basic picture of the cooling process in a lake is obtained with a 1D model (see [2])

$$\rho_w c_w \frac{\partial T}{\partial t} = \frac{\partial}{\partial z}\left(k_w \frac{\partial T}{\partial z} + Q_s\right) \tag{7.1}$$

where c_w is the specific heat of water, k_w is the thermal conductivity of water, and Q_s is the solar irradiance. The flux boundary conditions are $k_w \frac{\partial T}{\partial z} = K_0 + K_1[T_a - T(0)]$ at the surface, where the right side is the linearized air–water heat flux (see Sect. 3.5.3), and zero flux below the mixed layer. At ice formation, bottom heat flux is very small in deep lakes but can be 10–15 W m^{-2} in shallow lakes [7]. Assuming well-mixed conditions and absorbing net solar radiation in the surface flux (K_0), this integrates into

$$\frac{d}{dt}\tilde{T} = \frac{K_1}{\rho_w c_w H}\left(T_a - \tilde{T}\right) + \frac{K_0}{\rho_w c_w H} \tag{7.2}$$

where H is the mixed layer depth, and $\tilde{T}$ is the mixed layer temperature. Equation (7.2) is known as the slab model. For linear atmospheric cooling, $T_a = T_f - at$, the freezing time is

$$t_F = \frac{\rho_w c_w}{K_1} H + \frac{1}{a}\frac{K_0}{K_1} \tag{7.3}$$

The first term on the right-hand side is a lag due to the depth of the lake or mixed layer, and the second term represents the radiation balance effect.

Example 7.1 Equation (7.3) can be expressed as $t_F \approx C_1 H + C_0$. $C_1 \approx 2 - 3$ day m^{-1} or each metre in depth adds a 2–3-day delay to freezing; C_0 may have either sign, if the atmosphere cools by *a* ~5 °C month^{-1}, roughly, $|C_0| < 18$ days. In practice C_0 depends on the latitude via solar radiation climate, positive (negative) in low (very high) latitudes [8].

In cooling, there are two differences between lakes and rivers. First, continuous river flow needs an advection term, but here we assume that in the time scale of cooling the temperature is uniform along the flow and advection can be ignored. Second, rivers possess open spots in rapid streams where frazil[1] ice forms in cold weather. Frazil crystals move downstream and form a solid ice cover only when the buoyancy of ice crystals overcomes the turbulent energy production of the river flow. Thus, the predicted freezing date may refer to the start of frazil ice production or formation of near-shore border ice. River ice has a major influence on the fluvial hydraulics and geomorphology of freezing rivers. It adds on the wetted perimeter length and the friction for the water flow, and drifting ice blocks build up into ice jams, which may cause floods when hindering the water flow to pass (Portrait 7.1).

7.2 Structure and Properties of Lake and River Ice

7.2.1 Crystal Structure and Stratification

Nucleation temperature is below the freezing point in inland waters. Primary nucleation is heterogeneous, with suspended particles acting as the crystallization seeds, and supercooling remains small. After the initial nucleation, the 'extra cold' is recovered, and temperature returns to the freezing point. The first ice layer is *primary ice* [9], which may initiate in three different ways: in calm water, turbulent flow, and snowfall.

In calm water, crystallization begins at a temperature of 0.5–1.5 K below the freezing point. Ice crystal lattices grow on the surface into needles or platelets with crystal *c*-axis vertical, perpendicular to the needle axis or platelet

[1] Frazil comes the French word fraisil, which refers to coal cinders.

Portrait 7.1 Howard T. Barnes (1873–1950) is a classic person in lake ice and river ice science. He was born in Woburn, Massachusetts, moved to Montreal, Québec with his family in 1879, and performed his studies and most of his research at McGill University in Montreal. After receiving his D.Sc. degree (1900), he joined the University's Physics Department. In 1908 he succeeded Ernest Rutherford as Macdonald Professor of Physics and in 1919 became chairman of the department. His particular interest was research on icebergs and in reducing ice on the St. Lawrence River, with great contributions to thermodynamics of lake ice and river ice and ice jams. He published two classic books in ice science, *Ice formation, with special reference to anchor ice and frazil* (1906, J. Wiley & sons) and *Ice engineering* (1928, Montreal Renouf Publishing Co.). (*Photograph* © The Royal Society, Picture Library)

plane, favored by the anisotropic thermal conductivity of ice crystals. In turbulent water, supercooling is less than 0.1 K with frazil ice forming. Michel [10] suggested that near lateral boundaries (shore or ice cover), turbulence is weak, and a steep surface temperature gradient can be set up to support nucleation, while Osterkamp [11] considered that ice or cold soil particles could initiate nucleation falling on the supercooled turbulent water. Finally, snowfall is very effective due to wide distribution of falling snow crystals to trigger secondary nucleation in the water. During heavy snowfall, the primary ice layer forms of congealed slush.

Secondary ice grows beneath the primary ice into the basin water, and *superimposed ice* grows on top from precipitation and flooded lake water. Ice layers have a characteristic structure of *macro-crystals*, or grains, which are multiple crystals with individual crystal platelets overlain together, acting optically like single crystals. These macro-crystals are classified for their size, shape, and *c*-axis orientation (Table 7.1). In general, the slower the ice grows the larger will the crystals be. The main classes of crystal shape are planar, granular, and columnar, and orientation of crystals can be random or aligned.

7.2.2 Stratigraphy of Lake and River Ice

The crystal structure types are listed in Table 7.2. The primary ice layer can be thin (P1, P2) or of finite thickness (P3, P4), and its texture depends on the prevailing weather and water flow conditions. In calm water, *quiet freezing* [13] takes place (P1 and P2), typically in a cold, clear night with thermal radiation losses from the water surface; crystal size varies depending on the concentration of nucleation seeds and the rate of heat loss, and *c*-axes are dominantly vertical. Freeze-up starts from the shoreline, where the water is shallow and cools fast, and a near-shore border ice zone is often formed.

Frazil ice (P3) is common in rivers [1]. Crystals are fine circular disks with random *c*-axes, and they follow the turbulent river flow. Nucleating on suspended particles, frazil ice is effective in capturing suspended impurities. Crystals form shuga and grease ice and agglomerate into porous flocs which collide with each other. When their buoyancy increases enough, the flocs float on the surface and round, forming pancake ice (e.g., [14]) and gradually consolidate into a solid ice layer. When ice formation is initiated by ice or

Table 7.1 Classification of ice crystals based on their size, shape, and orientation [9, 10]

Crystal size	Fine	<1 mm
	Medium	1–5 mm
	Large	5–20 mm
	Extra large	>20 mm
	Giant	Metre scale
Crystal shape	Needles, plates	
	Granular (linear dimensions of the same magnitude)	
	Columnar: elongated vertically	
Crystal c-axis orientation	Vertical	
	Horizontal (random/aligned in 2D)	
	Random in 3D	

Table 7.2 Lake and river ice stratigraphy and forms of ice [9, 12]. Surface hoar and anchor ice have been added here

Ice type	Code	Grain size	*c*-axes	Conditions
Primary ice	P1	≥ Large	Vertical	Calm, slow growth
	P2	≥ Medium	Vertical/random	Calm, rapid growth
	P3	Fine	Random	Turbulent (frazil ice)
	P4	≤ Medium	Random	Snowfall
Secondary ice				
Congelation ice	S1	≥ Large granular	Vertical	Beneath primary ice P1
	S2	≥ Large columnar	Horizontal	Beneath primary ice P2
	S3	≥Large	Horizontal, aligned	Perennial lake ice
Frazil ice	S4	≤ Medium	Random	Frazil ice transport
	S5	≤ Medium	Random	Drained frazil slush
Superimposed ice				
Snow-ice	T1	≤ Medium	Random	Wet snow
	T2	≥ Medium	Random	Wet snow
Frozen pond	T3	As P & S	As P & S	Surface pond
Surface hoar	T4	Fine–extra large	Random	Moisture deposition
Agglomerate ice	R1	All size	All variants	Agglomeration of ice blocks
Anchor ice	R2	Fine–extra large	Random	Freezing on bottom

snow nuclei precipitating onto water (P4), ice crystals remain small and randomly oriented. Ice crystals may also originate from the water body itself as frozen splash drops or by evaporation–deposition mechanism. Frazil may also attach the bottom of a shallow water body to form *anchor ice* [1, 15] which, when released, brings bottom sediments to the ice cover.

Secondary ice is normally congelation ice in lakes (S1, S2, S3) and has congelation ice–frazil ice multi-layer structure (S4, S5) in rivers. *Congelation ice* crystals grow downward from the ice bottom surface, and flowing *frazil ice* crystals may form a new layer to ice bottom [9, 13, 16–18]. In large lakes, semi-persistent open water spots could serve as potential sites for generation of frazil ice, but the present authors are not aware of publications on that.

In a static lake ice cover, crystal size of congelation ice increases with depth, and typically large or extra-large crystals are found beneath a 10–20 cm thick *transition layer*. If the *c*-axes of primary ice layer are vertical (P1, P2), congelation ice has giant macro-crystals (S1) with the vertical *c*–axis orientation ([13, 19]). Otherwise, congelation ice crystal axes turn horizontal with depth through a geometric selection process [20, 21], and the crystals are vertical columns with 0.5–5 cm diameter and 5–50 cm height. Columnar congelation ice possesses anisotropy

Fig. 7.3 Frost flowers on ice in Lake Vähä-Tahko, Finland, January 2014. The l size of individual flowers is 5–10 cm. (*Photograph* Jari Toivonen, printed by permission)

in the *c*-axis alignment and the crystal shape but usually this is ignored.

Superimposed ice includes *snow-ice,*[2] or frozen slush (T1, T2) [22, 23], frozen surface ponds (T3), and surface hoar deposition (T4) (Fig. 7.3). The common form is snow-ice, which has fine or medium size crystals, depending on the size of the parent snow crystals, and random *c*-axes. Snow-ice grows down from the top of a slush layer, and a slush sub-layer may be locked below the snow-ice layer and persist for several weeks [24]. Snow-ice inherits air inclusions from the snow layer and has lower density than congelation ice.

Agglomerate ice (R1) consists of broken ice pieces. Ice cover can be broken by winds, waves, and changes in water level elevation, and ice blocks may drift and pile up to ridges and ice jams. This is common in rivers and large lakes. Anchor ice blocks can sometimes be recognized from the captured bottom sediments on the surface, risen due to buoyancy overcoming the binding forces at the bottom.

Example 7.2 Ice formation in a medium size boreal, oligo-mesotrophic lake, Lake Pääjärvi in southern Finland (61°40′N, 25°08′E). The surface area is 13.4 km^2, and the mean depth is 15.3 m. Ice structure was studied in 20 consecutive years in connection with an annual field course. The annual maximum ice thickness ranged from 24 to 80 cm, and the fraction of snow-ice was within 0–90% of the total ice thickness. Secondary ice was S1 (mostly) or S2 type. In one winter out of four, a several centimetres thick, semi-persistent slush layer was observed below snow-ice layer, with living algae. The statistics were (in cm):

Thickness statistic	Ice total	Congelation ice	Snow ice	Snow	Freeboard
Mean	44.3	31.0	13.3	7.5	2.5
St. dev.	12.5	12.5	9.1	8.0	0.9
Min	24	23	0	0	−5.0
Max	80	70	40	30	5.0

7.2.3 Snow Cover on Ice

Apart from arid and cold climate regions, such as the Tibetan Plateau and McMurdo Dry Valleys, ice in inland waters is covered with snow for a part of the season. The snow layer is soft, excellent insulator and has high albedo and low light transmittance. In metamorphic processes, liquid water beneath the ice is a heat source that supports depth hoar formation in cold weather, and a slush layer forms when snow is mixed with liquid water. In very cold conditions, adhesion between ice and snow is weak, and wind can clear the snow away if the snow accumulation keeps on a modest level. This is typical, e.g., in Lake Baikal.

Thickness of snow on ice is limited by flooding. Archimedes law gives the freeboard as

[2] Snow-ice is also called *white ice* due to high scattering of sunlight and whitish appearance, and congelation ice is called *black ice* due its dark appearance.

$$h_f = \left(1 - \frac{\rho_i}{\rho_w}\right)h_i - \frac{\rho_s}{\rho_w}h_s \qquad (7.4)$$

where h_i, ρ_i and h_s, ρ_s are the thicknesses and densities of ice and snow, respectively. Ice surface sinks down with snow accumulation. At $h_f < 0$, it is below the water surface level, and flooding may take place. For $\rho_s = 250$ kg m^{-3}, $h_s > \frac{\rho_w - \rho_i}{\rho_s} h_i = 0.33 \cdot h_i$ for the flooding condition. Water floods on ice through eventual cracks, and, mixed with snow, forms slush which turns into snow-ice in cold weather. However, cold freshwater ice may remain solid under the snow pressure, and then a new drillhole acts like an artesian well with water bursting up.

Snow is densified when mixed with water, and slush contains about half-and-half snow crystals and liquid water. High mountains with heavy snow accumulation, such as Sierra Nevada of California, introduce an extreme case. The ice cover consists of slush and frozen layers and the total thickness of this massif can be 3 m [25]. Slush can also be formed as a mixture of snow and rain or snow meltwater.

7.2.4 Impurities in Ice Cover

Lake and river ice cover contain impurities between ice crystals and in pores: gas bubbles, liquid inclusions, and particles, which are called *ice sediments* (Fig. 7.4). They originate from the water body, bottom sediments, and atmospheric deposition, and they influence on ice properties, particularly electromagnetic properties. They are examined by extracting gas from ice samples and analysing the meltwater of samples in laboratory [26, 27]. Impurities are transported with drifting ice and released to the water body mostly during the short melting period.

Gases are released from water body when it warms since the saturation level of dissolved gases decreases with increasing temperature. Bottom sediments also release gases, in particular methane, due to biogeochemical processes. Gas bubbles accumulate beneath ice cover and are locked inside during ice growth (e.g., [16]). Their size is usually 0.1–10 mm, but methane bubbles can reach several centimetres [28]. Considering 1% as the scale of the gas content in congelation ice, 910 kg m^{-3} is its convenient reference density. The gas (air) volume of snow-ice is 3–6% (e.g., [23]), with the reference density of 860–890 kg m^{-3}. Gas bubbles cause scattering of electromagnetic waves that is a major issue in light transfer through ice and in remote sensing.

Dissolved substances are segregated at freezing with only a small fraction (~1/10) captured in freshwater ice [26, 29]. Their concentration is in new ice $C_i = kC_w$, where C_w is the concentration in water, $k = k(C_w, \dot{h})$ is the segregation coefficient, and $\dot{h}$ is the growth rate of ice [29]. During ice growth, the surface water layer is enriched with dissolved substances, while ice melting has an opposite effect [6, 30]. In highly humic lakes, liquid humus inclusions have been found in ice [31]. Humus absorbs light at short wavelengths, as seen in light transmittance spectra [32].

The bulk concentration of sediment particles is similar in lake water and ice, but river ice can be enriched with sediments due to frazil ice [26]. Algae may grow in semi-persistent slush layers in snow-ice. In very shallow waters, freezing can take a large water fraction, and

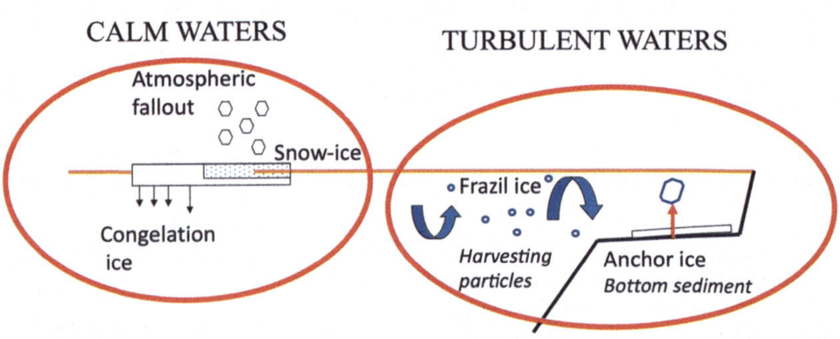

Fig. 7.4 The sources of ice impurities. Atmospheric deposition and flooded water in snow-ice formation are the main sources for lake ice, and for river ice, the main input is due to harvesting suspended particles by frazil ice and anchor ice formation

concentrations of harmful substances can become large in the remaining liquid water body [27]. Freezing reaching the bottom becomes a way to catch sediments in ice if the ice is lifted by rising water level.

7.3 Growth and Decay of Lake Ice

7.3.1 Congelation Ice Growth

Congelation ice growth is known as the Stefan's problem[3] or the diffusion equation with a moving boundary [33]. Ice can grow when the surface is colder than bottom which is at the freezing point. The latent heat released in freezing is conducted through ice to the atmosphere. The equation of heat conduction (Eq. 2.13) and the boundary conditions at the surface ($z = 0$) and at the bottom ($z = -h_i$) are:

$$\rho_i c_i \frac{\partial T}{\partial t} = \frac{\partial}{\partial z}\left(k_i \frac{\partial T}{\partial z} + Q_{s+} e^{\kappa_i z} \right) \quad (7.5a)$$

$$z = 0 : k_i \left.\frac{\partial T}{\partial z}\right|_{z=0^-} = \rho L_f \frac{dh}{dt} + Q_0 \quad (7.5b)$$

$$z = -h_i : \begin{cases} T = T_f \\ \rho_i c_i \dfrac{dh_i}{dt} + Q_w = -k_i \left.\dfrac{\partial T}{\partial z}\right|_{z=h_i^+} \end{cases} \quad (7.5c)$$

where the subscript i refers to ice, Q_{s+} is the part of solar radiation penetrating the near-surface layer, and $Q_w \geq 0$ is the heat flux from lake water to ice bottom. In length scales >10 m, the conduction time scale is more than a year (see Sect. 2.3). Thus, the growth of seasonal ice cover is a local process.

The time scale of vertical heat diffusion in ice, $\mathcal{T} = \frac{k_i}{\rho_i c_i} h_i^2$, overlaps the synoptic time scale (days to 1 week) when $h_i \sim \frac{1}{2}$ m that has motivated quasi-steady thermodynamic models where the left-hand side of Eq. (7.5a) is zero. When solar radiation is ignored, and conductivity is fixed, the quasi-steady approach gives a linear temperature profile (Fig. 7.5). For analytic analysis, the surface flux is taken as a linear form (Eq. 3.33) $Q_0 = K_0 + K_1(T_a - T_0)$, and we have a pair of equations for ice growth and surface temperature (e.g., [2]):

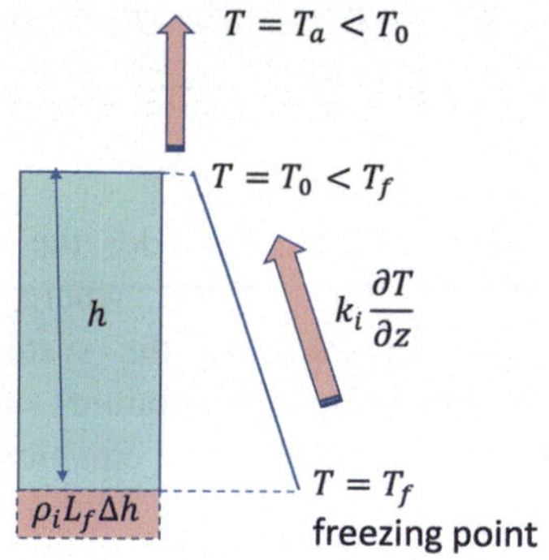

Fig. 7.5 The basis of quasi-steady models: heat release in ice growth is conducted through ice to atmospheric surface layer. Heat conduction takes place from warm to cold

$$\rho_i L_f \frac{dh_i}{dt} + Q_w = k_i \frac{T_f - T_0}{h_i} = -[K_0 + K_1(T_a - T_0)] \quad (7.6)$$

where $T_0 < T_f$ is the surface temperature. The conductive flux is inversely proportional to ice thickness that gives a 'self-insulation' effect.

Neglecting the heat flux from water and eliminating T_0, integration leads to

$$h_i(t) = \sqrt{(h_0 + h_a)^2 + a^2 S_R} - h_a \quad (7.7)$$

$$a = \sqrt{\frac{2k_i}{\rho_i L_f}},$$

$$S_R = \int_0^t \max(T_f - T_a - T_R, 0\,°\mathrm{C})\, d\tau, \quad T_R = \frac{K_0}{K_1},$$

$$h_a = \frac{k_i}{K_1}$$

where S_R is the sum of *freezing-degree-days*, $a \approx 3.4$ cm(°C day)$^{-1/2}$ is the freezing-degree-day coefficient, T_R is a shift term due to the radiation balance, h_a~10 cm is buffering by the atmospheric surface layer, and h_0 is the initial thickness. Letting $K_1 \to \infty$, we have $T_0 = T_a$ and the Stefan [33] formula $h_i = \sqrt{h_0^2 + a^2 S_a}$,

[3] Josef Stefan (1835–1893), a Slovenian physicist, one of the leading scientists in the history of thermodynamics.

and letting $K_0 \to 0$ we have $T_R = 0$ and the Barnes [34] formula $h_i = \sqrt{(h_0 + h_a)^2 + a^2 S_a} - h_a$.

Insulation by snow cover adds the heat flux through snow, $k_s(T_{0s} - T_0)/h_s$, where the subscript s refers to snow, to the system (7.6). Eliminating the surface temperatures of ice and snow, we obtain the differential equation

$$\rho_i L_f \frac{dh_i}{dt} + Q_w = k_i \frac{\max(T_f - T_a - T_R, 0\,^\circ\mathrm{C})}{h_i + h_a + \beta h_s} \quad (7.8)$$

where $\beta = k_i/k_s \sim 10$. Clearly $\beta h_s \gg h_a$ as soon as $h_s > 5$ cm. Equation (7.8) cannot be solved analytically, even with $Q_w = 0$, since the thickness and conductivity of snow depend on time. There is an equilibrium solution for fixed forcing where the heat flux from water is conducted through a fixed ice and snow cover to air:

$$h_i + h_a + \beta h_s = \frac{k_i(T_f - T_a - T_R)}{Q_w} \quad (7.9)$$

For an acceptable equilibrium, $h_i > 0$.

7.3.2 Snow-Ice Growth

Snow-ice grows down from the upper surface of a slush layer. The heat transfer system is Eq. (7.8) with $Q_w = 0$ and the latent heat released reduced by the liquid water fraction or the porosity (ν) of slush [2, 35]:

$$\rho_i L_f \nu \frac{dh_{si}}{dt} = k_i \frac{\max(T_f - T_a - T_R, 0\,^\circ\mathrm{C})}{h_{si} + h_a + \beta h_s} \quad (7.10)$$

where h_{si} is the snow-ice thickness. Snow cover limits the growth, but with $\nu \sim ½$ much less heat needs to be conducted. A new flooding event produces slush on top of the existing snow-ice, and thus it is possible that alternating slush and snow-ice layers exist.

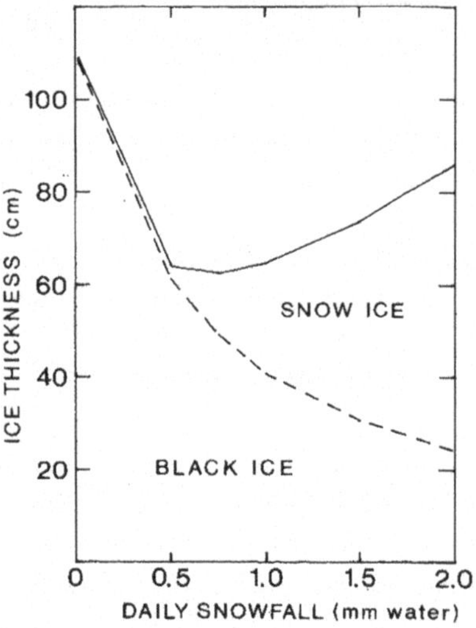

Fig. 7.6 Thickness of congelation ice and snow-ice- vs. daily snow accumulation at latitude 65°N in Finland. Air temperature follows the climatological daily means [35]

Insulation and snow-ice formation have opposite impacts by snow to total ice thickness, i.e., more snow accumulation can either increase or decrease ice thickness (Fig. 7.6). For fixed freezing-degree-days, the maximum ice thickness is obtained when the ice is bare. In case of the strongest insulation, when no slush is yet formed, ice thickness is about half of the bare ice case, and when snow accumulation is more, ice thickness increases due to snow-ice formation reaching asymptotically 0.7–0.8 times the bare ice case [35].

7.3.3 Lake Ice Deterioration

Deterioration of lake ice progresses in a similar manner for all ice types. Depending on albedo and light attenuation coefficient [36], melting takes place at the top and bottom surfaces and inside the ice (Fig. 7.7, see [37–40]). Lateral surface patchiness develops with snow, bare ice, wet snow or ice, and open water spots. The patchiness develops due to positive feedback between melt rate and albedo. Heat flux from lake water to ice increases in the melting period since a fraction

Fig. 7.7 Decay of a lake ice cover is a chaotic, driven by solar radiation, air temperature, precipitation, and wind. The result is a patchy ice cover during the decay period. (*Photograph* Matti Leppäranta)

of under-ice solar heating melts ice. Near the shoreline and in shallow areas, water beneath ice warms fast and ice disappears first.

Melting ice is described by thickness and porosity which together give the volume of ice per unit area as $V = (1 - \nu)h_i$ [39]. Ice temperature is at the melting point. Surface melting and bottom melting are obtained from the boundary conditions, and internal melting is integrated from the solar radiation:

$$\frac{dh_i}{dt} = -\frac{1}{1-\nu} \cdot \left[\frac{\max(Q_0, 0) + Q_w}{\rho_i L_f}\right] \quad (7.11a)$$

$$\frac{d\nu}{dt} = \frac{Q_{s+}}{\rho_i L_f h_i}\left(1 - e^{-\kappa_i h_i}\right) \quad (7.11b)$$

Internal melting begins from the crystal boundaries, and columnar-grained ice appears then as *candled ice*. Increasing porosity weakens the ice, and at $\nu = \nu^* \sim 0.5$, the ice collapses under its own weight. Broken ice exposes more surface area resulting in a fast final stage of melting.

Example 7.3 Taking the net solar flux in the surface (K_0), we have

$$\frac{dV}{dt} = -\frac{1}{\rho_i L_f}\left\{\max\left[K_0 + K_1(T_a - T_f), 0\right] + Q_w\right\}$$

The common degree-day melting model is $\frac{dh}{dt} = -A \cdot \max[(T_a - T_f), 0\,°\mathrm{C}], A \sim 0.5\ \mathrm{cm}\ (°\mathrm{C\ day})^{-1}$. The corresponding degree-day coefficient is here $\frac{K_1}{\rho_i L_f} \sim 0.3\ \mathrm{cm}\ (°\mathrm{C\ day})^{-1}$. This difference reflects the role of the radiation in melting.

Spatial thickness inhomogeneities arise in lake ice, associated with convection under ice. At extreme, heat flux from water keeps permanent openings in ice cover, as known from open geothermal lakes in cold climate regions. Local vertical circulation due to, e.g., groundwater inflow or gas flux from the bottom sediments, can create semi-persistent holes, up to tens of centimetres across, which are dangerous for on-ice walking (Fig. 7.8). The condition for persistent opening is $Q_w + K_0 + K_1(T_a - T_f) \geq 0$. Historical photographs have revealed cell-like structures in water body, marked with the radial or star-shaped melt patterns or even melt holes, considered to result from Rayleigh-Bénard convection [41]. Ice melting from below resulting from stationary

Fig. 7.8 A semi-persistent opening (size ~1/2 m) forced by vertical circulation in the lake water body, March 2021, Lake Oulujärvi, Finland. (*Photograph* Katja Huotari. Printed by permission)

giant rings (kilometres) was discovered in Lakes Baikal and Hovsgol [42].

7.4 River Ice Growth and Decay

7.4.1 Ice Formation

River ice forms in static manner, as in lakes, or in dynamic manner with frazil ice generation in streams (see Sect. 7.2; [10, 43]). There is a large interannual variation in ice conditions depending on the weather and river flow. Frazil ice production is in balance with the surface heat loss. In rapid streams, the surface may remain open even through the winter. These spots act as 'ice factories' during cold periods, producing frazil ice crystals which flow downstream.

Heat flux from water to ice bottom is modeled similarly with the air–ice turbulent flux,

$$Q_w = \rho_w c_w C_{Hw} (T_w - T_f) |u_w - u_i| \quad (7.12)$$

where $C_{Hw} \sim 10^{-3}$ is the water–ice heat exchange coefficient, T_w is the water temperature, and u_w and u_i are the velocities of water and ice, respectively. The magnitude of this heat flux is tens of watts per square metre in winter and more in ice decay period [44]. As a result, ice thickness in rapid streams can be much less than in low-speed sections. This is usually not visible and forms a major risk factor for on-ice river crossing.

7.4.2 Frazil Ice

Frazil ice crystals flow free in turbulent eddies and are transported by the mean current. Their production can be fast in cold weather, given by

$$\rho_i L_f \frac{dh_F}{dt} = -\left[K_0 + K_1(T_a - T_f) + \left(1 - e^{-\kappa_w H}\right)Q_{s+} + Q_w\right] \quad (7.13)$$

where h_F is the frazil volume of per unit area or equivalent thickness, and H is the river depth. This equation can be directly integrated since the right side is independent of h_F. There is no self-insulation feedback as in congelation ice growth. Equation (7.13) works when $h_F \geq 0$.

Replacing T_f by T_0, the equilibrium $T_0 \geq T_f$ represents open surface but $T_0 < T_f$ is not a possible equilibrium. For example, in the presence of strong geothermal heat flux or solar radiation of 200 W m^{-2}, a freshwater river or lake may stay open down to the air temperatures of about -10 ° C.

Example 7.4 Take $Q_0 = K_0 + K_1(T_a - T_f)$. In high-latitude winter, $K_0 \sim -50$ W m^{-2} and $K_1 \sim 15$ W m^{-2}. For $T_a - T_f = -15$ °C, we have $Q_0 \sim -275$ W m^{-2} that corresponds to frazil production by 7.9 cm day^{-1}. Congelation ice growth is close to that in the first day, 7.3 cm day^{-1}, but in the fifth day it has reduced to 3.1 cm d^{-1}. Adding the heat flux $Q_w = 50$ W m^{-2}, frazil production is 6.5 cm day^{-1}, and congelation ice grows first by 1.4 cm day^{-1}, on the fifth day by 0.5 cm day^{-1}, and the equilibrium thickness is 50 cm.

Frazil ice deposits on the bottom of solid ice cover, riverbed, man-made objects like boat anchors or fishing nets, and on upstream side of structures such as bridge piers. This occurs where the river slope changes from steep to mild. Deposition of frazil crystals causes problems at water intakes in hydropower stations where frazil accumulation may block the water flow. Ice crystals accumulate into flocs and form slush and pancake ice on the surface. Pancakes may freeze together to form large floes, and then this accumulating ice layer progresses upstream. In strong flow, pancakes coming to the edge will submerge and continue drifting downstream. The layer thickens until the yield strength is high enough to resist the pressure by the water flow. Thick ice needs long time to break up, therefore thick sections are potential ice jam locations.

7.4.3 Breakup and Jams

There are two types of ice jams, freeze-up and breakup jams (Fig. 7.9). The former is less common and occurs in early winter. Freeze-up jams increase the river stage and may cause flooding, but flash floods are unlikely. Frazil ice dynamics may initiate ice jams, and when the frazil mass thickens, the potential danger of ice jams increases. Breakup jams in spring are more likely to create flash floods. Many plains and wetland ecosystems require periodic flooding for their replenishment of moisture, sediment, and nutrients.

Normally river ice breakup process is thermal decay with a series of mechanical events. In pure thermal decay, ice melts in place, and there are no ice jams. River flow transports heat from open spots to increase the rate of melting. Mechanical breakup is more complicated. A breakup event begins when the forcing by the flow and ice pressure overcomes the strength of the ice cover. Increasing river discharge, resulting from increasing spring runoff or liquid precipitation, increases the hydrodynamic forcing that can trigger a mechanical breakup. Another factor to add the river flow can be a release from a jam breakage upstream. When a static ice cover is broken due to an increase in the river discharge, the consequent raising river stage can loosen the ice from the riverbanks and move it downstream, broken into small floes. In places the floes may pile up into an ice jam. The blocked river is enlarged, and backwater effects may cause floods upstream while the downstream flow decreases. These floods can extend fast, and large chunks of ice can cause damage.

The potential for mechanical breakup is tied to water level elevation at the freeze-up. A low water level at freeze-up results in a greater probability for a breakup at relatively low flows. On the other hand, if a river freezes at high water, the

Fig. 7.9 An ice breakup jam in River Kyrönjoki, Finland, March 31, 2021. (*Photograph* South Ostrabotnia Centre for Economic Development, Transport and Environment. Printed with permission)

level is expected to drop in winter as the discharge decreases which needs to be recovered before the ice cover can start drifting. Hinge cracks along the sides of a river indicate that the middle of the ice cover has sagged due to decreases in discharge. A sudden development of an ice jam blocks the flow and results in a dramatic stage that can be associated with a mechanical breakup upstream.

The locations of ice jams are largely determined by the river morphometry, and therefore jam build-ups often occur, year-by-year, at specific, known locations. The key morphometry factors are the shape of cross-sections, depth, and slope. Where the slope decreases, frazil ice is deposited, ice cover is thick, and occurrence of an ice jam is more likely. Sediment deposits change the river profile that impacts on the ice jam potential.

Ice jams and flooding can pose threats to communities along the river basins. The formation of an ice jam can be a rapid, catastrophic process, and its prediction is difficult. Ice jam floods have been reported in Europe, e.g., in Tornionjoki River at the border between Finland and Sweden, and Oder River at the border between Germany and Poland. River ice models have been developed for ice jamming and the consequent flooding. Factors that affect ice cover breakup can be monitored in real time, including discharge, weather (particularly air temperature and precipitation), snow accumulation in the drainage basin, ice strength, and ice thickness.

7.5 Ice Mechanics

7.5.1 Mechanics Questions

Ice mechanics problems concern ice forces, bearing capacity, and ice drift. Congelation ice of freshwater basins largely agrees for its mechanical properties with freshwater ice used in laboratory experiments (see Sect. 2.2). In the melting period, ice cover becomes porous with decreasing strength and loses contact with the land boundaries. Then the ice may then be pushed to move by winds and currents, riding up or piling up on shores. The ice decay period contains the highest risks for offshore and shoreline constructions.

When a lake ice cover reaches critical thickness, it spans a solid, stable lid, and the ice is safe to go until the beginning of melting period. The stability length scale of ice cover is $\ell \propto h_i^2$ [45]. At h_i~10 cm, we have ℓ~5 km. In very large lakes, the required thickness is large, and ice mechanical displacements can take place all winter. This question is treated in detail in Chap. 8 (Sea Ice).

***Example** 7.7 'Ice quakes'.* In January 2009, a rare 'ice quake' event was experienced on the shore of Lake Vesijärvi, Finland. Ice thickness was about 10 cm, and strong northerly wind blew across the lake. Residents in new, high buildings about 40 m from the shoreline were alarmed by vibrations. The frequencies were 2–10 Hz at the floors, around the lowest natural frequency of the houses in the so-called pendel vibration mode [46]. The houses were supported by vertical concrete piles on stiff moraine, and the site was filled in the lake with a 6 m thick filling resting on original quite soft 15 m thick clay/silt layer above the moraine. The in-lake building site was regarded as the main reason why the vibrations due to the cracking lake ice could be transported to the buildings via the soil. When ice melting began in spring, vibrations started again after a 2-month quiet period. The highest measured velocities were 1.2 mm s^{-1}, exceeding the comfort limit of the residents, but no structural damages were observed.

7.5.2 Ice Forces

Lake and river ice is a polycrystalline continuum, and its mechanical properties have a large variability due to large crystal size, high homologous temperature, and impurities (e.g., [1, 10]). Isotropy and homogeneity are normally assumed, but anisotropy may exist in the crystal structure, and homogeneity can be violated by a vertical temperature gradient. In situ field tests have the great advantage loading in natural conditions. Maximum loading takes place when the ice fails at a structure. This force depends on ice properties, failure mode, and the structure itself. Loading cases are concerned with ships, harbors, piles, docks, and other shoreline constructions, and the natural shoreline itself (Fig. 7.10). The failure is not necessarily achieved due to short fetches with forces below the maximum.

The fundamental method to evaluate ice loads is the Korzhavin [47] equation for the load at ice failure on vertical structures:

$$F = Ik_1k_2\sigma bh_i \tag{7.14}$$

where I is the indentation factor, k_1, k_2, and b are the contact factor, shape factor, and width of the structure, respectively, and σ is the ice strength. The indentation factor is due to the 3D nature of ice stress around the structure, $I = I(\delta)$, $\delta = b/h_i$, with $I \approx 2.5$ for narrow structures and $I \approx 1$ for wide structures. The parameters k_1 and k_2 represent the contact between the structure and ice, $k_1 \geq ½$ for the incomplete contact between the ice edge and structure and $k_2 \leq 1$ for the form of the structure cross-section. In case of a non-vertical structure, the inclination of the loaded surface has also influence.

Further simplifications have been derived from Korzhavin [47] model. The force on a pier-type structure is [1]

$$F = K\sigma_c bh_i \tag{7.15}$$

where σ_c is the compressive strength of ice, and K is a scaling coefficient, which depends on the mechanical properties of ice and the geometry of the pile structure. For a narrow structure, the aspect ratio is an important parameter, $K = 2.5$ for $\delta = 1$ and $K \to 1$ for $\delta \gg 1$. For wide structures and wedge-shaped piles, $K = 1$ can be taken. For an inclined structure, $K < 1$ depends on the slope angle of the structure and ice–structure friction. When the structure is frozen to the ice, changes in water level elevation cause vertical forces. The largest forces result in fast changes, since then the ice behaves in elastic manner.

Fig. 7.10 Offshore outcome in ice breakup in Kemijoki River in spring 1930. Site is at Valmarinniemi, 5 km upstream from the river mouth. *Finnish Heritage Agency, photographer unknown*

Example 7.6 Consider a 1 m wide pile in freshwater ice with $\sigma_c = 10$ MPa. For thin ice (10 cm), $\delta = 10$ and $K \approx 1$, and the maximum load is $F = 1$ MN; for thick ice (1 m), $\delta = 1$ and $K \approx 2.5$, and the maximum load is $F = 25$ MN

Example *7.7 Thermal pressure.* In Lake Mälaren, Sweden thermal pressure has caused damage in the past [48]. In winter 1953, a swing bridge was partly damaged. Air temperature rose by 14 °C in 12 h, and 22 cm thick ice expanded and pressed towards the bridge. The forcing was estimated as 270 kN m^{-1}. In winter 1976, four beacons mounted on pile groups were destroyed by thermal expansion, with pressures estimated as 100–270 kN m^{-1}.

7.5.3 Bearing Capacity of Ice

The bearing capacity of lake ice and river ice is usually sufficient for traffic use (see [49, 50]). The practical solution of this problem was known in prehistoric times, shown by the wide winter traffic networks. Olaus Magnus Gothus [51] described ice crossing by skiing, skating (metal and bone blades), riding and sledges in North Europe. Battles on lake ice are known from the Medieval times (Fig. 7.11).

The physical basis of the bearing capacity was presented by Hertz [52]. Substantial progress in the science and technology of lake ice roads was made in Soviet Union during the Siege of Leningrad in 1941–1944. Ice road over Lake Ladoga, known as 'The road of life', was the only connection outside the city, and the bearing capacity was deeply worked on (see [49]). The formulation of the theory by Wyman [53] has been widely used in the western literature as the historical source.

Consider a one-dimensional ice beam on water foundation under a concentrated load P at the origin. The ice is bent down while supported by water pressure $\rho_w g w$, where w is the deflection (Fig. 7.12). Thus, liquid water body acts mathematically as an elastic foundation, i.e., the response of the foundation is proportional to deflection, and (e.g., [1])

Fig. 7.11 Battle on ice between Novgorod and Livonian Knights on the ice of Lake Peipsi/Chudskoe in 1242. A snapshot from the film Alexander Nevsky (1938) by Sergey Eisenstein, one of the famous scenes in the film history. According to the legend, the ice broke under the Livonian troops

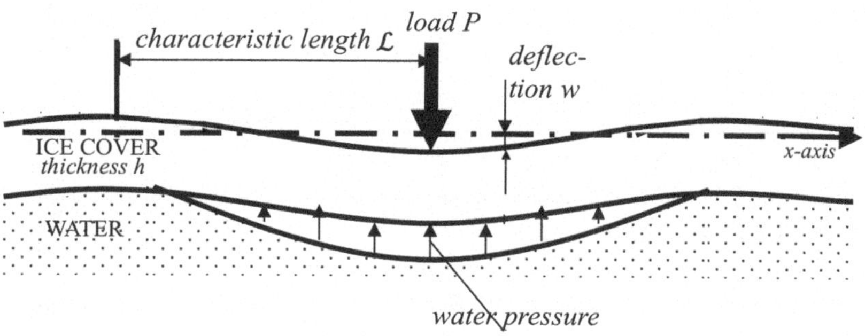

Fig. 7.12 Bending of ice beam of thickness h under load P; $\mathcal{L}$ is the characteristic length, and w is the deflection [49]

$$EJ\frac{d^4w}{dx^4} = -\rho_w g w, \quad J = \frac{bh_i^3}{12} \tag{7.16}$$

where E is elastic modulus, J is moment of inertia of the beam, and b is the beam width. The load P is accounted for in the boundary conditions. Under static load, the solution is

$$w(x) = -\frac{P}{2\rho_w g \mathcal{L}_1^2} e^{-\lambda x}[\sin(\lambda x) + \cos(\lambda x)],$$
$$\lambda = \frac{1}{\mathcal{L}_1} = \sqrt[4]{\frac{\rho_w g}{4EJ}} \tag{7.17}$$

for $x \geq 0$ (symmetric for $x \leq 0$). Deflection forms a wave exponentially damping with distance from the origin; λ is the wave number, and its inverse $\mathcal{L}_1$ is the characteristic length. With the elastic properties of lake ice (Table 2.1), $\mathcal{L}_1 = 4.2$ m for $h_i = 0.1$ m. For 100 kg load (P= 981 N m^{-2}), deflection at the origin is $w(0)$= 2.9 mm, much less than freeboard (8.3 mm). The maximum tensile stress occurs at the bottom surface at the origin.

In two dimensions, the problem, studied in polar coordinates (r, φ), is reduced to the Bessel equation, and the solution involves special functions (see [49, 50]). The 2D characteristic length is $\mathcal{L} = \sqrt[4]{\frac{Eh_i^3}{12\rho_w g(1-\nu^2)}}$, which represents the size of the deflection bowl or action radius. The deflection at $r = 0$ and the stress at $r \to 0$ are

$$w(0) = \frac{P}{8\rho_w g \mathcal{L}^2} \tag{7.18a}$$

$$\sigma_r = \sigma_\varphi = \frac{3(1+\mu)}{2\pi}\frac{P}{h_i^2}\log\left[\left(2\frac{\mathcal{L}}{r}\right) - c\right] \tag{7.18b}$$

where $c = 0.577216$ is Euler's constant. This stress becomes singular at $r = 0$ so that a finite size of the loading area needs to be considered. Equation (7.18b) has been the basis of practical formula of the bearing capacity showing that the stress scales as P/h_i^2, and $\mathcal{L}$ represents the radius of influence of a point load, $\mathcal{L} \approx 16\, h^{3/4}\ \mathrm{m}^{1/4}$ works for freshwater ice [54].

Deflection forms an exponentially damping wave bowl. At $r \gg \mathcal{L}$, it is practically invisible, and ice plate can be considered infinite with respect to the vertical load. Within the region $r < \mathcal{L}$, the upper surface of the ice is pressed down. The deflection bowl is symmetric at the beginning of loading at small deflections. At the bearing capacity limit, first radial cracks appear at the lower surface under the ice load. The solution of distributed loads is obtained by superposition of the solutions of concentrated loads.

Interest in dynamical loads originates from loads due to a moving vehicle along an ice road [55]. There is a critical vehicle speed, which can lead to the fracture of ice cover due to resonance of the vehicle motion and the shallow water wave motion in the water caused by the moving pressure [56]. The wave speed is $c_s = \sqrt{gH}$, where H is the water depth. If the moving vehicle has the critical speed, the resonance strengthens the bending of ice, and, in theory, the bearing capacity disappears. At subcritical speeds, maximal bending moments occur under load, while at supercritical speeds, the peak is in front of the load.

Study Problems

1. Solve Eq. (7.2) when $T_a = -at$ and K_0 and K_1 are constant. What would be realistic values for a, K_0 and K_1?
2. (a) Derive the solution of ice growth (7.7) from Eq. (7.6); (b) For air temperature –5 °C, determine the ice thickness evolution over 100 days using Stefan and Barnes formulas. When the relative and absolute difference have their maxima?
3. Assume albedo and light attenuation coefficient are 0.5 and 1 m^{-1} for ice, and 0.9 and 10 m^{-1} for snow. For solar radiation of 500 W m^{-2}, evaluate the ice and snow thickness combinations when primary production is possible, i.e., irradiance at least 20 W m^{-2}, under snow-covered ice.
4. (a) How thick ice must be to carry a person, whose weight is 80 kg? (b) How much ice will bend under a car for weight of 1000 kg when ice thickness is 25 cm. What is the radius of influence of the car? (c) How shallow a lake must be for a runner to achieve the critical speed in dynamic ice loading?
5. The heat flux is given by $Q = K(T_a - T_0)$, where $K = 2{\cdot}10^{-3}$ W °C^{-1} m^{-2} is the heat transfer coefficient. Estimate the frazil ice production and congelation ice growth in 10 days under the air temperature of –20 °C.

References

1. Ashton G (ed) (1986) River and lake ice engineering. Water Resources Publications, Littleton
2. Leppäranta M (2023) Freezing of lakes and the evolution of their ice cover, 2nd edn. Springer, Berlin
3. Magnuson J, Lathrop R (2014) Lake ice: winter, beauty, value, changes, and a threatened future. Lake Line 43:18–27
4. Bates RE, Bilello MA (1966) Defining the cold regions of the Northern Hemisphere. Technical report 178. US Army Cold Regions Research and Engineering Laboratory, Hanover
5. Moreira S, Schultze M, Rahn K, Boehrer B (2016) A practical approach to lake water density from electrical conductivity and temperature. Hydrol Earth Syst Sci 20:2975–2986
6. Kirillin G, Leppäranta M, Terzhevik A, Granin N, Bernhardt J, Engelhardt C, Efremova T, Golosov S, Palshin N, Sherstyankin P, Zdorovennova G, Zdorovennov R (2012) Physics of seasonally ice-covered lakes: a review. Aquat Sci 74:659–682
7. Zdorovennova G, Terzhevik A, Palshin N, Efremova T, Bogdanov S, Zdorovennov R (2021) Seasonal change in heat flux at the water-bottom sediment boundary in a small lake. J Phys Conf Ser 2131: 032080

8. Leppäranta M, Wen L (2022) Ice phenology in Eurasian lakes over spatial location and altitude. Water 14: 1037
9. Michel B, Ramseier RO (1971) Classification of river and lake ice. Can Geotech J 8(36):36–45
10. Michel B (1978) Ice mechanics. Laval University Press, Quebec
11. Osterkamp TE (1977) Frazil ice nucleation by mass-exchange processes at the air-water interface. J Glaciol 19(81):619–625
12. Gow AJ (1986) Orientation textures in ice sheets of quietly frozen lakes. J Cryst Growth 74:247–258
13. Gow AJ, Govoni JW (1983) Ice growth on post pond, 1973–1982. CRREL report 83–4. U.S. Army Cold Regions Research and Engineering Laboratory, Hanover, NH
14. Nghiem SV, Leshkevich GA (2007) Satellite SAR remote sensing of Great Lakes ice cover, part 1. Ice backscatter signatures at C band. J Great Lakes Res 33: 722–735
15. Shen HT (2006) A trip through the life of river ice—research progress and needs. In: Proceedings of the 18th IAHR international symposium on ice, pp 85–91
16. Li Z, Huang WF, Jia Q, Leppäranta M (2011) Distributions of crystals and gas bubbles in reservoir ice during growth period. Water Sci Eng 4(2):204–211
17. Shumskii PA (1956) Principles of structural glaciology [in Russian]. English transl. 1964 by Dover Publications
18. Sokol'nikov VM (1957) Some regularities in the formation and growth of ice cover: case study of Lake Baikal. Tr Baikal'Skoi Limnol St AN SSSR 15:58–64
19. Leppäranta M, Kosloff P (2000) The thickness and structure of Lake Pääjärvi ice. Geophysica 36(1–2): 233–248
20. Müller-Stoffels M, Langhorne PJ, Petrich C, Kempema EW (2009) Preferred crystal orientation in freshwater ice. Cold Reg Sci Technol 56:1–9
21. Pounder E (1965) Physics of ice. Pergamon Press, Oxford
22. Adams WP, Roulet NT (1980) Illustration of the roles of snow in the evolution of the winter cover of a lake. Arctic 33(1):100–116
23. Palosuo E (1965) Frozen slush on lake ice. Geophysica 9(2):131–147
24. Leppäranta M (2009) A two-phase model for thermodynamics of floating ice. In: Proceedings of the 6th workshop on baltic sea ice climate, report series in geophysics, vol 61, pp 146–154. Department of Physics, University of Helsinki, Helsinki
25. Melack JM, Sadro S, Sickman JO, Dozier J (2021) Lakes and watersheds in the Sierra Nevada of California: responses to environmental change. University of California Press, Berkeley
26. Leppäranta M, Tikkanen M, Virkanen J (2003) Observations of ice impurities in some Finnish lakes. Proc Estonian Acad Sci Chem 52(2):59–75
27. Yang F, Li C, Leppäranta M, Shi X, Zhao S, Zhang C (2016) Notable increases in nutrient concentrations in a shallow lake during seasonal ice growth. Water Sci Technol 74(12):2773–2883
28. Wik M, Crill PM, Bastviken D, Danielsson Å, Norbäck E (2011) Bubbles trapped in arctic lake ice: potential implications for methane emissions. J Geophys Res 116:G03044
29. Weeks WF, Ackley SF (1986) The growth, structure and properties of sea ice. In: Untersteiner N (ed) Geophysics of sea ice. Plenum Press, New York, pp 9–164
30. Huang W, Zhang Z, Li Z, Leppäranta M, Arvola L, Song S, Huotari J, Lin Z (2021) Under-ice dissolved oxygen and metabolism dynamics in a shallow lake: the critical role of ice and snow. Water Resour Res 57(5)
31. Salonen K, Leppäranta M, Viljanen M, Gulati R (2009) Perspectives in winter limnology: closing the annual cycle of freezing lakes. Aquat Ecol 43(3): 609–616
32. Lei R, Leppäranta M, Erm A, Jaatinen E, Pärn O (2011) Field investigations of apparent optical properties of ice cover in Finnish and Estonian lakes in winter 2009. Estonian J Earth Sci 60(1):50–64
33. Stefan J (1891) Über die Theorie der Eisbildung, insbesondere über Eisbildung im Polarmeere. Ann Phys, 3rd Ser 42:269–286
34. Barnes HT (1928) Ice engineering. Renouf Publishing Company, Montreal
35. Leppäranta M (1983) A growth model for black ice, snow ice and snow thickness in subarctic basins. Nord Hydrol 14(2):59–70
36. Warren SG (2019) Optical properties of ice and snow. Philos Trans R Soc A377:20180161
37. Jakkila J, Leppäranta M, Kawamura T, Shirasawa K, Salonen K (2009) Radiation transfer and heat budget during the melting season in Lake Pääjärvi. Aquat Ecol 43(3):681–692
38. Leppäranta M, Lindgren E, Arvola L (2016) Heat balance of supraglacial lakes in the western Dronning Maud Land. Ann Glaciol 57(72):39–46
39. Leppäranta M, Lindgren E, Wen L, Kirillin G (2019) Ice cover decay and heat balance in Lake Kilpisjärvi in Arctic tundra. J Limnol 78(2):163–175
40. Zhang Y, Fregona M, Loehr J, Ala-könni J, Song S, Leppäranta M, Li Z (2023) A field study on ice melting and breakup in a boreal lake, Pääjärvi, in Finland. Cryosphere 17:2045–2058
41. Mironov D, Terzhevik A, Kirillin G, Jonas T, Malm J, Framer D (2002) Radiatively driven convection in ice-covered lakes: observations, scaling, and a mixed layer model. J Geophys Res 107(C4):3032
42. Kouraev AV, Zakharova EA, Rémy F, Kostianoy AG, Shimaraev MN, Hall NMJ, Suknev AY (2016) Giant ice rings on lakes Baikal and Hovsgol: inventory, associated water structure and potential formation mechanism. Limnol Oceanogr 61:1001–1014
43. Martin S (1981) Frazil ice in rivers and oceans. Annu Rev Fluid Mech 13:379–397

44. Marsh P, Prowse TD (1987) Water temperature and heat flux at the base of river ice covers. Cold Reg Sci Technol 14(1):33–50
45. Leppäranta M (2013) Land-ice interaction in the Baltic Sea. Estonian J Earth Sci 62(1):2–14
46. Makkonen L, Törnqvist J, Kuutti J (2010) Vibrations of buildings induced by thermal cracking of lake ice. In: Proceedings of the 20th IAHR ice symposium
47. Korzhavin KN (1962) Vozdeystviye l'da na inzhenertrye sooruzheniya [The action of ice on engineering structures]. Izdatel'stvo Sibirskogo Otdel, Akademiya Nauk SSSR, Novosibirsk
48. Bergdahl L, Wernersson L (1978) Calculated and expected thermal ice pressure in five Swedish lakes. Department of Hydraulics, Chalmers University of Technology, report series B:7
49. Goldstein R, Leppäranta M, Onishchenko D, Osipenko N (2017) Bearing capacity of ice cover. UNESCO EOLLS Encyclopedia
50. Masterson DM (2009) State of the art of ice bearing capacity and ice construction. Cold Reg Sci Technol 58:99–112
51. Gothus MO (1555) Description of the northern peoples, Venice, Italy [in Latin]
52. Hertz H (1884) Über das Gleichgewicht schimmender elastischen Platten. Wiedmanns Annalen der Physik und Chemie 22:449–455
53. Wyman M (1950) Deflections of an infinite plate. Can J Res A28:293–302
54. Gold L (1971) Use of ice covers for transportation. Can Geotech J 8:170–181
55. Bregman GR, Proskuryakov BV (1943) Ice crossing-Sverdlovsk. Gidrometeoizdat
56. Nevel DE (1970) Moving loads on a floating ice sheet. CRREL research report 261. Cold Regions Research and Engineering Laboratory, Hanover, NH

8 Sea Ice

In summer in the Arctic Ocean. The artist Gunnar Brusewitz (1924–2004) is well known for his bird drawings. This picture is from the Swedish Ymer-80 expedition in the European sector of the Arctic Ocean in the summer of 1980. (Reproduced from Gunnar Brusewitz, Arktis sommar (1980), with permission from Wahlström & Widstrand, Stockholm)

In this chapter, the physics of sea ice is presented. The characteristics specific to sea ice are saline ice and large length scale. Salinity has a major influence on the physical properties of sea ice, and the large length scale is reflected in the broken structure and drift of sea ice fields. The first section introduces freezing seas, followed by sea ice structure from microscale to large scale in Sect. 8.2. Section 8.3 is about sea ice thermodynamics, and Sects. 8.4 and 8.5 present drift ice for its morphology and dynamics. Mathematical models are highly developed and applied in ice

M. Leppäranta, P. Uotila, *Geophysics of the Cryosphere*, Springer Textbooks in Earth Sciences, Geography and Environment, https://doi.org/10.1007/978-3-031-93071-3_8

engineering, marine forecasting, and climate research.

8.1 Introduction

Sea ice occurs in about 10% of the world's ocean surface, growing, melting, and drifting under the influence of solar, atmospheric, oceanic, and tidal forcing [1–4]. Most sea ice lies in the Arctic and Antarctic seas above 60° latitude (Table 8.1). Moreover, seasonally freezing marine basins exist in lower latitudes, such as the Baltic Sea and the Hudson Bay and, even below 40°N, the Bo Hai Sea, China. Growing sea ice captures salts from the parent seawater, and salt inclusions act as the key small-scale characteristics with significant influence on the properties of sea ice [3, 5]. A 'sea ice landscape' consists of leads and ice floes with ridges, hummocks, and other forms of pressure ice (Fig. 8.1). Drifting of sea ice is typically a few kilometres per day, and in the polar oceans, sea ice moves slowly towards lower latitudes limiting the age of most sea ice to not more than a few years.

Dissolved salts are important in freezing of seawater and sea ice properties (e.g., [3, 6]). The density standard of seawater is TEOS-10 (Thermodynamic Equation of Seawater), but the Unesco standard is often used for simplicity (see Annex A1.4). The temperatures of the freezing point T_f and the maximum density T_m are depressed with salinity S (see Annex A1.4). Polar seawater usually has a salinity of 30–35‰ (per mille), with the freezing point between −1.6 and −2.0 °C. The temperature of maximum density decreases faster than the freezing point, and $T_m = T_f$ at $S = 24.7‰$. For $S > 24.7‰$ free convection continues until freezing, but otherwise, a stable stratification forms at $T < T_m$, as in lakes.

People have lived on the coast of freezing seas for a long time, which has affected their everyday lives ([7, 8]). Seal hunting, whaling, and fishing have been based on sea ice, and sea ice cover has served for traffic and communication. Ice bridges have formed across narrow straits and between islands, providing traffic routes (Fig. 8.2). For shipping, sea ice has acted as a barrier, and in modern times, forces on structures such as oil and gas platforms and wind farms by drifting ice have become important research topics, as well as the transport and dispersion of pollutants and oil spills in ice-covered waters. Since the 1990s, polar environment, ecology, and climate change have become central themes in sea ice research [9].

The history of sea ice science has roots in the 1800s [3, 10]. Karl Weyprecht led the Austro-Hungarian Expedition on the schooner *Tegetthoff* to Franz-Josef Land in 1872–1874 and examined sea ice thickness and morphology [11]. He found from empirical data that sea ice growth is proportional to the square root of the freezing-degree-days, explained by Stefan [12] using the Fourier heat conduction law. Fridtjof Nansen drifted with the topsail schooner *Fram*, moored to ice, across the Eurasian Arctic Ocean in 1893–1896. The wind-driven drift speed was, on average, 2% of the surface wind speed, and the drift direction deviated 30° to the right from the wind direction [13], which was explained by Ekman [14] with his planetary boundary layer model. In the Southern Ocean, the German ship *Deutschland* drifted

Table 8.1 Sea ice in the polar oceans

Polar ice	Arctic ocean	Southern ocean
Maximum ice area	12 million km^2	19 million km^2
Minimum ice area	5 million km^2	3 million km^2
Renewal time	2 years	1 year
Thickness	1–4 m	½ to 2 m
Ice type	Mostly congelation ice	Mostly frazil ice
Ice morphology	Much ridged ice	Pancake ice typical
Land ice	Ice islands only	Large, tabular icebergs

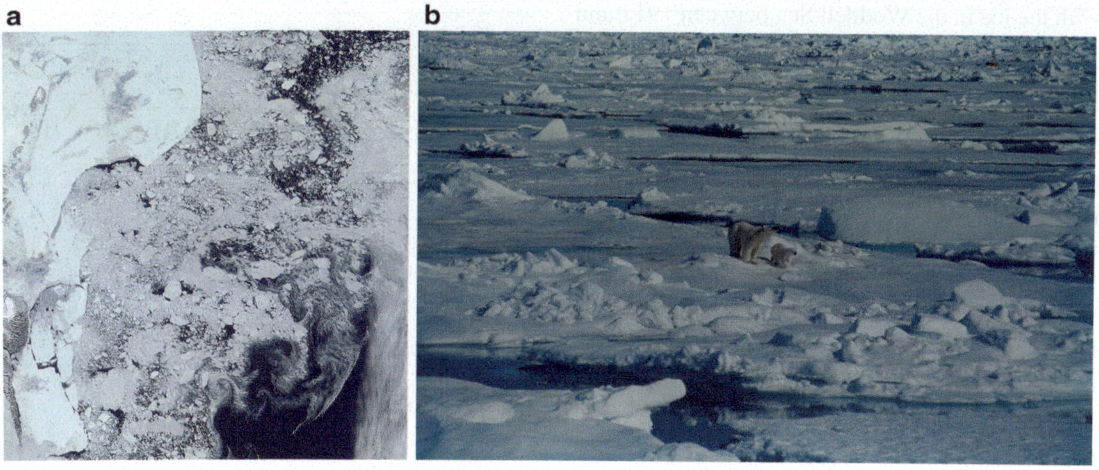

Fig. 8.1 (**a**) Drift ice in the Greenland Sea. (**a**) A section of Aqua–MODIS image on August 2, 2015. (*Source* NASA Earth Observatory). (**b**) Surface photograph from the Greenland Sea, July 1983. (*Photograph* Matti Leppäranta)

Fig. 8.2 Ice road across the northern Baltic Sea from Vaasa, Finland to Umeå, Sweden. The road was used in severe winters until 1972, when all-winter shipping was commenced in the region. *Finnish Heritage Agency*. (*Photograph* Rafael Olin)

with the ice in the Weddell Sea between 1911 and 1912 [15]. Antarctic sea ice was studied relatively little until winter cruises commenced in the 1980s [16].

The Arctic and Antarctic Research Institute (AARI) was established in 1920 in Petrograd (later Leningrad, presently St. Petersburg), focusing on the Northern Sea Route. A series of North Pole drifting stations commenced in 1937, where scientific campsites drifted from the Pole to the Greenland Sea in 1 year. The pioneering book *Arctic sea ice* by Zubov [4] was published. North American scientists worked on *T-3* or *Fletcher's ice island* in the Beaufort Sea between 1952 and 1978. When the Beaufort Sea oil resources were found, the AIDJEX (Arctic Ice Dynamics Joint Experiment) programme was performed during the 1970s, producing the basis of the present understanding of sea ice dynamics [17, 18].

Winter navigation in the seasonal sea ice zone was the background to developing sea ice charting and forecasting methods in the subarctic freezing seas, e.g., the Baltic Sea [19, 20]. In Japan, a research programme was begun in the 1940s on the drift of ice in the Sea of Okhotsk [21], leading to the establishment of a coastal radar system on the north coast of Hokkaido for sea ice monitoring [22].

8.2 Sea Ice Structure

Sea ice is examined over a wide range of scales. The microscale includes individual grains and ice impurities extending to 0.1 m. In the local scale, 0.1–10 m, sea ice is a solid polycrystalline continuum with a sub-structure classified according to the ice formation mechanisms. Ice floe scale extends from 10 m to 10 km, including individual floes and ice forms such as rubble, pressure ridges, and landfast ice. In mesoscale (10^2 km) and large scale (10^3 km), sea ice medium appears granular, with ice blocks and floes as the 'grains', called drift ice.

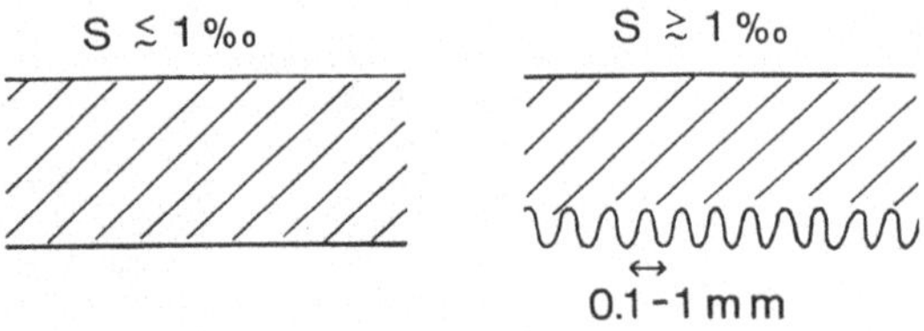

Fig. 8.3 Ice–water interface is planar in low-saline water ($S \leq 1$‰) and cellular in saline ($S \geq 1$‰) water, limit approximate and case dependent. Segregation of dissolved substances in ice growth is effective with planar but limited with cellular interface

8.2.1 Microscale

Sea ice crystals are built of water molecules, and dissolved salts in seawater are rejected from the crystal lattice. A cellular ice–water interface forms (Fig. 8.3) due to constitutional supercooling caused by the difference in the molecular diffusion of salt and heat in liquid seawater [23]. A large fraction of salts can be captured in the cells. Sea ice crystal boundaries are jagged compared with smooth freshwater ice crystals, and the bottom layer of sea ice is the fragile *skeleton layer*, which contains a large fraction of brine. The threshold salinity of the parent water for the congelation ice type is 1–2‰ [24, 25].

8.2.2 Local Scale

Sea ice grows through different mechanisms, which is observed in its layered vertical structure. The main forms are *congelation ice*, *snow-ice*, and *frazil ice* [26] as in freshwater basins (see Sect. 7.2). Congelation ice is the dominant form in the Arctic Ocean, while frazil ice, produced in leads and polynyas in turbulent open water [27], is dominant in the Antarctic seas where it is also the source of pancake ice. Snow-ice is common in the seasonal sea ice zone, with large snow accumulation [28]. *Anchor ice* crystallizes in the shallow sea bottom in open sea, turbulent conditions. As observed in the Siberian shelf, ice lumps may rise to the surface with bottom

Fig. 8.4 Platelet ice at the bottom of ice cover in Saroma-ko lagoon, north coast of Hokkaido in the Sea of Okhotsk. The scale numbers show centimetres. (*Photograph* by Matti Leppäranta)

sediments due to buoyancy. *Platelet ice*, large platelet crystals, forms of upwelling supercooled water at the bottom of sea ice. It was first found in Antarctica, where floating ice shelves provide the rising supercooled water [29]. Upwelling glacial meltwater becomes supercooled due to decreasing pressure. This form has also been observed in a saline lagoon where river inflow is supercooled by the cold lagoon water beneath landfast sea ice (Fig. 8.4).

The salinity of sea ice refers to the salinity of its meltwater. Salts are stored in sea ice in liquid brine pockets and salt crystals depending on the temperature [30]. Individual salts have their crystallization temperature or eutectic temperature. The composition of sea ice can be determined from the phase diagram (Fig. 8.5). The salinity of brine, S_b, corresponds to the freezing point at the ambient temperature T, $T_f(S_b) = T$, S_b increasing with decreasing temperature. A temperature change is always associated with freezing or melting at brine pocket boundaries.

Mathematical formulae for the fractions of brine and crystallized salts are presented in Cox and Weeks [31] for $T \leq -2$ °C and by Leppäranta and Manninen [32] for -2 °C $< T < 0$ °C. At temperatures higher than -8.2 °C, in practice all salts are in brine, and the brine volume is

$$\nu_b = \frac{\rho_i S_i}{\rho_b S_b} \tag{8.1}$$

where ρ_i is the sea ice density, S_i is the sea ice salinity, and ρ_b (kg m^{-3}) $\approx 1000 + 0.82 \cdot S_b$‰ is the density of brine. At lower temperatures, the fraction of salt crystals becomes important, especially below -22.9 °C, which is the eutectic temperature of sodium chloride [31]. Brine volume is the principal factor that influences the properties of sea ice.

The density of sea ice is $\rho = (1 - \nu_a - \nu_b)\rho_i + \nu_b\rho_b$ where ν_a is the gas content [33],

Example 8.1 For $S_i = 10$‰, at $T = -8$ °C, we have $S_b \approx 127$‰ and $\nu_b \approx 0.07$; and at $T = -4$ °C, we have $S_b \approx 71$‰ and $\nu_b \approx 0.13$. The brine volume increases in a nonlinear manner and, asymptotically $\rho_i S_i \to \rho_b S_b$, $\nu_b \to 1$ as $T \to 0$ °C. In practice, sea ice collapses when the brine volume exceeds ½. Though the freezing point of standard sea water is -2 °C, sea ice crystal lattice is pure ice and can exist at higher temperatures. Many semiempirical formulae exist in the literature for physical properties of sea ice with singularity as $T \to 0$ °C that needs reconsideration when studying sea ice warmer than -2 °C.

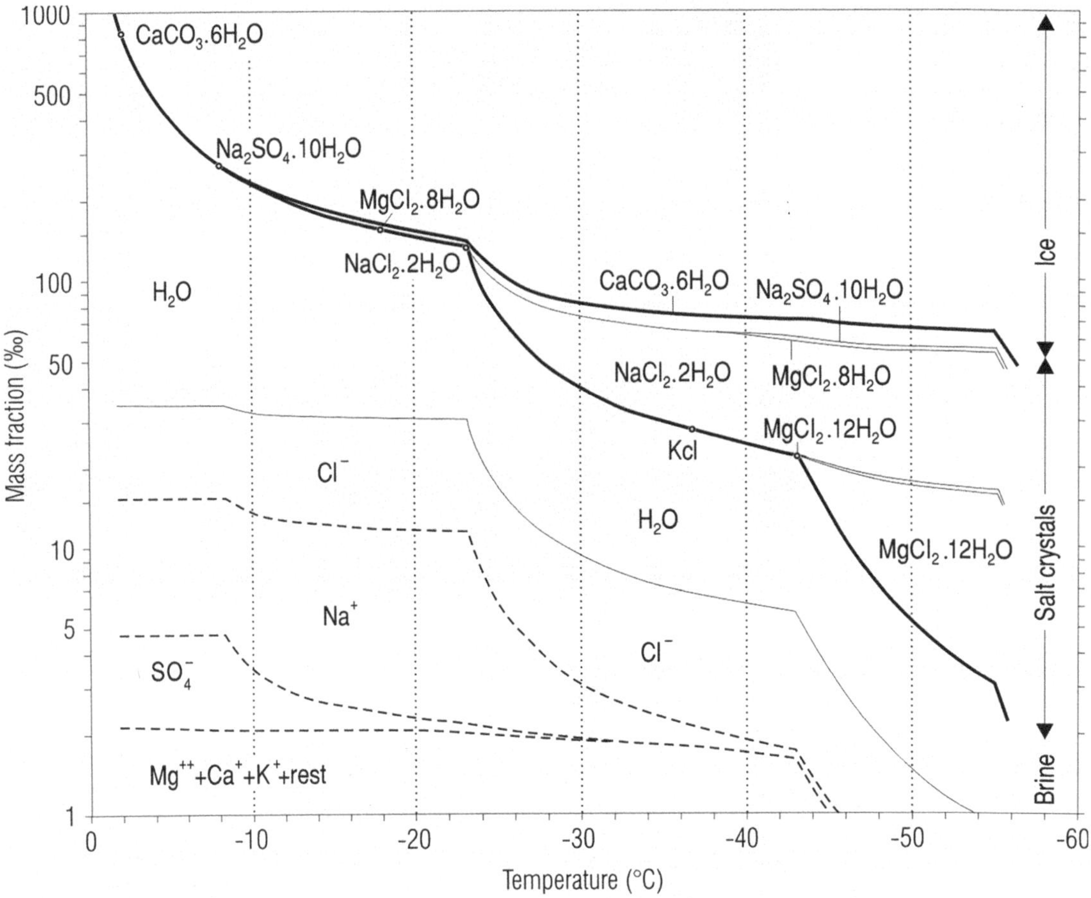

Fig. 8.5 Sea ice phase diagram with the proportions of ice, brine, and solid salts as a function of temperature. Different salts begin to crystallise at their eutectic points. Warm sea ice has a large fraction of brine while salt crystals are abundant in very cold ice. (Redrawn from Assur [30])

The salinity of new sea ice is a fraction of 0.25–0.5 of the salinity of the parent water, depending on the growth rate. Salinity decreases with time as the brine flows down. The most effective loss mechanism is gravity drainage and flushing in summer when brine pockets enlarge in warm ice into a drainage network. Multi-year sea ice has an equilibrium salinity of about 1‰, with summer loss compensating for winter gain. Brine pockets can serve as habitats for biota with algae captured in the freeze-up process [5, 34]. The skeleton layer may become coloured brown or green by algae.

8.2.3 Ice Floe Scale

Sea ice appears in various visible *ice types* based on the appearance, i.e., how the ice looks to an observer on a ship or in an aircraft. They originate from practical shipping in ice-covered waters, and they have been standardized by the World Meteorological Organization (WMO), published in sea ice nomenclature ([35], first ed. 1970), with updated versions available on the Internet (Table 8.2). The polar oceans also contain *ice of land origin*. By the calving mechanism, pieces of ice break away from land ice masses facing the

Table 8.2 Common ice types [35]

New ice Grease ice Nilas	Recently formed ice less than 10 cm Frazil crystals have coagulated to a soupy layer on the surface Thin, elastic film that bends under the action of swell
Young ice	Thicker than new ice, 10–30 cm
First-year ice	Thicker than young ice, up to 2 m
Multi-year ice	Sea ice survived at least one summer, thickness > 2 m. [also: old ice]
Linear openings Crack Fracture Lead	Opened due to mechanics Less wide than 1 m Wider than crack, up to kilometres long Fracture navigable by surface vessels
Polynya Coastal polynya Offshore polynya	Persistent opening enclosed in ice Driven by steady offshore winds Driven largely by the upwelling of warm, deep water
Pancake ice	Circular pieces of new ice or young ice, ~1 m across, often raised rims
Brash ice	Accumulations of floating ice made up of blocks not more than 2 m across
Level ice Landfast ice	Unaffected by mechanical deformation *Sea ice* which forms and remains fast along the coast
Deformed ice Hummocked ice Ridge Rafted ice	Ice types created by pressure [also: pressure ice] Pieces of ice are piled haphazardly to form an uneven surface Wall of broken ice forced up by pressure, upper (lower) part is *sail* (*keel*) One floe overrides another
Grounded ice	Aground ice in shallow water [also: stamukha, grounded ridge]

ocean. They are classified according to their size: *icebergs* (top more than 5 m above sea level), *ice islands* (top about 5 m above sea level and area more than a few thousand square metres), *bergy bits* (top 1–5 m above sea level and area 100–300 m^2), and *growlers* (smaller than bergy bits). These pieces differ from sea ice floes by their ice quality (freshwater ice), thickness and three-dimensional character.

Sea ice ridges are the thickest form of deformed ice (Fig. 8.6). Their thickness is typically 5–30 m, and over large areas, their volume may account for up to one-half of the total ice volume [1, 3, 36]. The ratio of keel depth to sail height is 4–5 for first-year ridges and about 3 for multi-year ridges, and a typical sail height is 1–2 m in the Central Arctic. In ice engineering, ridges are of deep concern because they put the highest ice loads on structures within first-year sea ice [37], scour the sea bottom [38], and influence on-ice traffic conditions. The record ridge is from the Beaufort Sea, with a sail height of 12 m and a keel depth of 45 m [39].

8.2.4 Drift Ice

In large basins, apart from landfast ice, solid sea ice lids are statically unstable and break into fields of ice floes, forming a granular medium, *drift ice*. In scales more than 10 km, drift ice can be approximated as a continuum (Portrait 8.1). The mechanical behaviour of this system depends on the sea ice concentration (Table 8.3) and thickness (e.g., [1, 40, 41]). The concentration tells of the mobility of drift ice, and it can be determined by satellite remote sensing (e.g., [42]).

8.3 Sea Ice Thermodynamics

8.3.1 Theory

Thermal properties of sea ice depend on temperature and salinity [43, 44]. Sea ice has no definite melting point, as melting is involved when ice warms to dilute the brine and vice versa, which is reflected in its specific heat (Fig. 8.7a). Brine is a

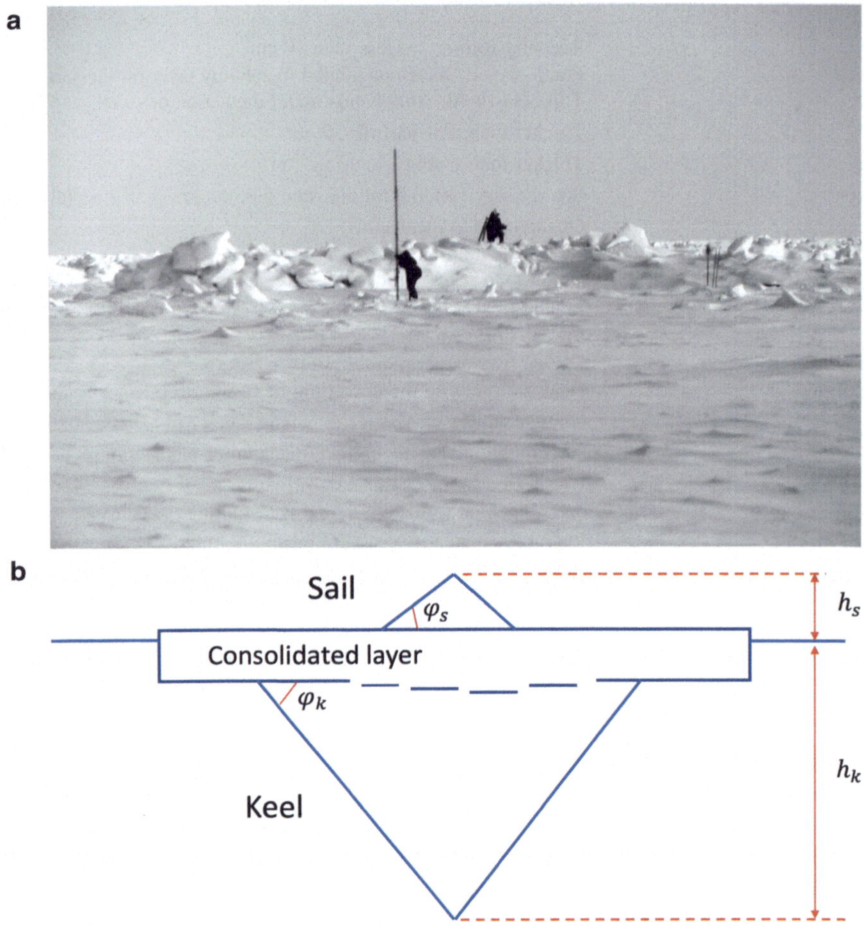

Fig. 8.6 A sea ice pressure ridge: (**a**) a field photograph, and (**b**) an idealised cross-sectional diagram. Sail and keel are characterized by their heights and slopes, and at the sea surface level there is a consolidated layer. (*Photograph* Matti Leppäranta)

little weaker heat conductor than ice, and therefore the sea ice thermal conductivity is smaller than the freshwater ice thermal conductivity (Fig. 8.7b). Brine dynamics delays the cooling and warming of sea ice compared with freshwater ice, but the maximum annual ice thickness is not much affected. The thermodynamics of congelation ice and snow-ice can be solved using similar one-dimensional models following the lake ice (Sect. 7.3).

The surface heat balance of sea ice follows the general form in Sect. 3.5 (e.g., [45, 46]). The temperature and salinity of sea ice are coupled, and the full solution of sea ice thickness and temperature needs them both [47]. However, a semiempirical model often prescribes or solves the salinity distribution. The thermal conduction law for a sea ice layer is (see Eq. 2.13)

$$\rho c\frac{\partial T}{\partial t} = \frac{\partial}{\partial z}\left(k\frac{\partial T}{\partial z} + Q_{s+}e^{\kappa z}\right) \tag{8.2a}$$

$$z = 0 : k\frac{\partial T}{\partial z} = \rho L_f\frac{dh_i}{dt} + Q_0 \tag{8.2b}$$

$$z = -h_i : \begin{cases} T = T_f \\ \rho L_f\dfrac{dh_i}{dt} + Q_w = -k\dfrac{\partial T}{\partial z} \end{cases} \tag{8.2c}$$

Portrait 8.1 Fridtjof Nansen (1861–1930), born in Christiania (presently Oslo), Norway, was one of the famous early polar scientists. He greatly contributed to the research of sea ice and polar oceanography, particularly through drifting with ice based on the ship *Fram* across the Eurasian Arctic Ocean in 1893–1896. In 1922, he was awarded the Nobel Peace Prize for his work on behalf of the displaced victims of World War I and introduced 'Nansen passport' for stateless persons. The photograph shows Fridtjof Nansen in August 1896 after returning from the Fram expedition. (*Photograph* Johan Wickstrøm, Nansen-basen, National Library of Norway)

where c, k, and κ are the specific heat, thermal conductivity, and light attenuation coefficient of sea ice, respectively, h_i is the ice thickness, Q_{s+} is the solar radiation penetrating to ice, Q_0 is the surface heat balance, and Q_w is the heat flux from seawater to ice. The bottom boundary condition could be refined by assuming a skeleton layer with ice density $\rho' < \rho$. Specific heat is one order of magnitude larger in warm than in cold ice, which slows thermal diffusion (see Example 2.2). Cooling warm ice takes time due to squeezing the brine pockets and vice versa, delaying temperature adjustment compared with freshwater ice [48].

Sea ice deterioration is a coupled thermal and mechanical process where ice thickness decreases and porosity increases. Ice loses its strength and breaks when the thickness is still finite. The final disappearance of ice is faster due to mechanical breakage and drift offshore than due to pure melting. Landfast ice thickness is primarily a thermodynamic problem but not only so [49, 50]. The Eurasian landfast ice zone is wide due to the shallow continental shelf, and congelation ice is the dominant ice type (e.g., [51]). In the annual cycle, landfast ice disappears in August–September due to melting, breakage, and drift, and after summer, new ice begins to grow in open water. Therefore, multi-year landfast ice is rare. In Greenland fjords, old, even 10 m thick landfast ice exists, known as *sikussak* in the local language [52]. There, the snow accumulation protects the ice from melting in summer, and offshore drift is limited.

8.3.2 Analytic Models

Analytical sea ice growth models are based on the quasi-steady solution of the heat conduction law with a moving lower boundary, the so-called Stefan's problem [53]. The models are as for freshwater ice (see Sect. 7.3). Quasi-steady models contain thermal conductivity but not specific heat. Therefore, the model parameters do not differ much between sea ice and freshwater ice cases.

In Arctic seas, oceanic heat flux is not intense, and using Eq. (7.7) with $K_0 = 0$, Zubov's [4]

Table 8.3 Classification of ice concentration or ice coverage A [35]

Verbal	Numerical	Verbal	Numerical
Ice free	$A = 0$	Very open drift ice	$0 < A < 30\%$
Open drift ice	$30\% \leq A < 50\%$	Close drift ice	$50\% \leq A < 70\%$
Very close drift ice	$70\% \leq A < 90\%$	Compact drift ice	$90\% \leq A \leq 100\%$

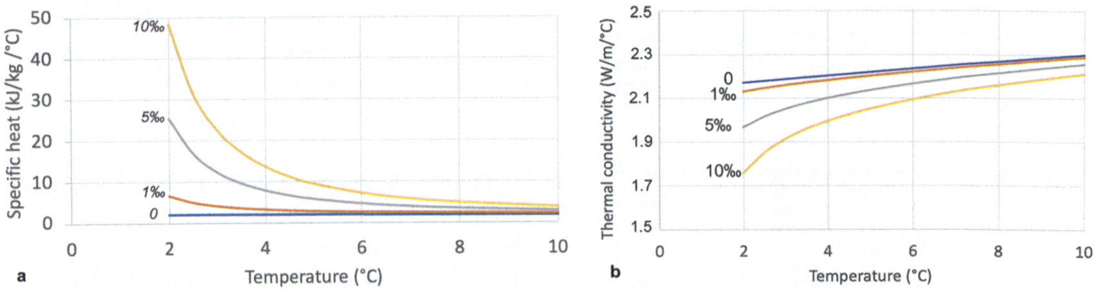

Fig. 8.7 Thermal properties of sea ice from −10 to −2 °C based on the formulas of Schwerdtfeger [44]. (**a**) Specific heat, and (**b**) thermal conductivity. The permille (‰) labels refer to the sea ice salinity. Below −10 °C differences are small

formula is obtained. This gives the thickness of congelation ice as

$$h_i = \sqrt{(h_0 + h_a)^2 + a^2 S_a} - h_a \tag{8.3}$$

where $a = \sqrt{2k/\rho L_f}$, h_0 is the initial thickness, $h_a \approx 10$ cm, and S_a is the freezing-degree-days of air temperature. At −5 °C, conductivity is about 20% lower in first-year sea ice than in freshwater ice. This reduces a by 10%, to $a \approx 3$ cm $(°C \cdot d)^{-1/2}$. On the Siberian coast, the freezing-degree-days can be 5000 °C day over a winter [49], and the first-year ice thickness is at the end of winter $h_1 = \sqrt{h_a^2 + a^2 S_a} - h_a \approx 2.0$ m.

The extent of the sea ice melting period depends mainly on the latitude, air temperature, and precipitation. During this period, ice is nearly isothermal. Ice thickness decreases, and porosity increases until mechanical breakage occurs, after which the final melting is fast. The surface patchiness due to the positive albedo feedback is typical of the melting period, with snow-covered spots, bare ice spots and melt ponds. In the polar seas, summer melting depends primarily on the radiation balance, accounting for a reduction of ice thickness by $\Delta h \sim 0.5 - 1.0$ m, which does not depend on the thickness of the ice.

The thickness of second-year ice is at the end of winter (from Eq. 8.3)

$$h_2 = \sqrt{(h_1 - \Delta h + h_a)^2 + a^2 S_a} - h_a \tag{8.4}$$

With $S_a = 5000$ °C·d and $\Delta h = 0.75$ m, we have $h_2 = 2.4$ m. Finally, the multi-year ice equilibrium thickness h_∞ is obtained from the condition $h_{n+1} = h_n$. Ignoring the minor impact of h_a, the equilibrium thickness is at the end of winter

$$h_\infty = \frac{h_1^2}{2\Delta h} + \frac{\Delta h}{2} \tag{8.5}$$

Thus, with $h_1 = 2.0$ m and $\Delta h = 0.75$ m, we have $h = 0$ m (Fig. 8.8): ice melts by 0.75 m in summer, and old ice grows back in winter. In Eq. (8.5), the first term on the right shows that multi-year ice thickness is highly sensitive to the first-year ice thickness and summer melt, as seen in recent decades in the Arctic.

In Antarctic seas, frazil ice formation is the dominant mechanism of sea ice growth [54]. Snow-ice formation is significant, and oceanic heat flux is intense. Frazil ice is connected to the open structure of the divergent Antarctic sea-ice cover with an open boundary at the polar front. The frazil growth law is written

$$\frac{dh_f}{dt} + Q_w = \frac{Q_{s+} + Q_0}{\rho L_f} \tag{8.6}$$

where h_f is the volume of frazil ice per area. Frazil crystals may drift downstream or join to form grease ice and pancake ice. When a solid ice cover has formed, the oceanic heat flux limits the growth, introducing an equilibrium ice thickness

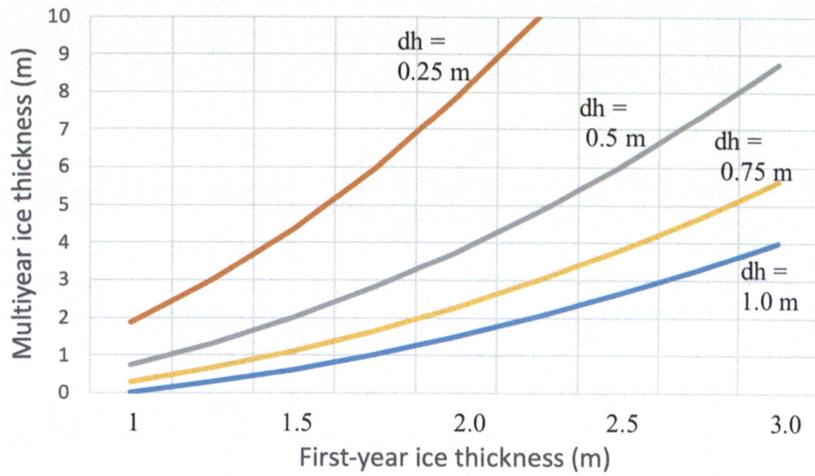

Fig. 8.8 Multi-year ice thickness (in metres) as a function of first-year ice thickness and summer melt based on analytic modelling

$$h + h_a = \frac{k_i(T_f - T_0)}{Q_w} \tag{8.7}$$

where the oceanic heat flux is exactly conducted through the ice cover.

8.3.3 Numerical Modelling

The first time-dependent numerical model on sea ice thermodynamics sea ice was presented by Maykut and Untersteiner [55]. It was a congelation ice model for the Arctic Ocean showing the multi-year ice equilibrium cycle with ice thickness of 3–4 m in the standard case. The most critical factors in the sensitivity study were albedo, oceanic heat flux, and snow accumulation. A simplified version was developed by Semtner [48] for climate research. Bitz and Lipscomb [43] presented an energy-conserving model accounting for internal brine pocket melting on surface ablation, and a snow-ice layer was added to the model by Leppäranta [56] and Saloranta [57]. Cheng et al. [58] included an advanced atmospheric boundary layer model to improve the accuracy of turbulent fluxes.

Numerical models solve the heat conduction Eqs. (8.2a, 8.2b, 8.2c) for the evolution of the temperature of snow and ice and ice thickness using a finite difference or finite element method [43, 47, 55, 59]. A model may have separate congelation ice, snow-ice, slush, and dry snow layers. The surface temperature is solved from the surface heat balance that couples the ice with the atmosphere. The internal layers interact: snow-ice growth depends on snow accumulation and congelation ice growth and vice versa.

For example, the annual landfast cycle in the New Siberian Islands was simulated using a time-dependent thermodynamic model forced by weather observations [49]. The standard reference model run yielded 167 cm maximum annual ice thickness (Fig. 8.9), slightly less than the maximum ice thickness measured in Sogo Bay at Tiksi, south of the New Siberian Islands [60]. The simulations showed that ice thickness could reach 220 cm without snow. The effect of precipitation on ice thickness is complicated since it depends on timing, and its phase is difficult to predict. Yang et al. [49] performed sensitivity studies for the summer decay of landfast ice, and a realistic range of about 1 month resulted in the date of ice breakup due to melting only.

8.4 Sea Ice Kinematics

8.4.1 Morphology of Drift Ice

A sea ice cover can be divided into zones in terms of dynamics [1, 61]: central pack, shear zone, landfast ice, and marginal ice zone. The central pack is free from immediate influence from the boundaries, while the shear zone and landfast ice

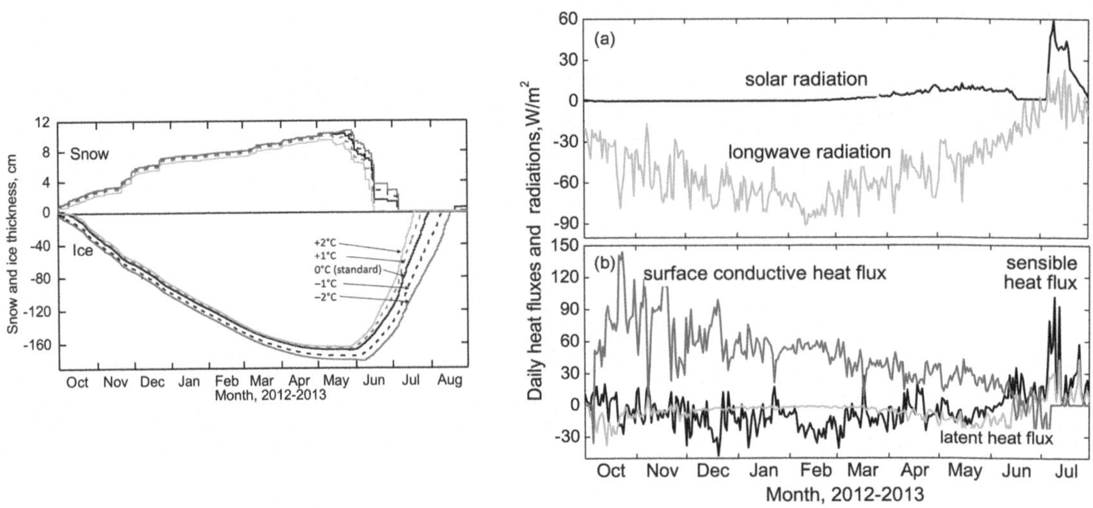

Fig. 8.9 Simulation of the landfast ice thickness in the East Siberian Sea, Left: The sensitivity of annual snow and ice thickness cycle to the changes of air temperature by ±1, ±2 °C. The zero level refers to the top surface of the ice. Right: The modelled surface heat fluxes (daily mean) for the standard experiment: (**a**) net radiation fluxes, and (**b**) turbulent heat fluxes and conductive heat flux from ice [49]

form the coastal boundary zone, and the marginal ice zone is the boundary zone towards the open ocean. Mapping of ice conditions is based mainly on satellite data, and the results are presented in ice charts [42, 62] (Fig. 8.10). Ice extent and concentration are well represented in ice charts, but ice thickness information is limited.

The landfast ice zone is next to the shoreline and extends to 10–20 m depths in Arctic seas [4, 63, 64]. Its width depends on the ice thickness, the sea bottom topography, and the areal density of islands and grounded ice forms, extending up to 200 km from the shoreline at the New Siberian Islands. It is stationary most of the time, and during the decay process, it may break and drift away [49]. In the shear zone, the ice mobility is restricted by the boundary geometry and friction; a well-developed shear zone is found in the Beaufort Sea (e.g., Hibler et al. [65]). The marginal ice zone (MIZ) is characterized as the drift ice zone, which 'feels the presence of open ocean' and extends to 100 km from the ice edge [66, 67]. This distance corresponds to the penetration distance of the ocean swell into a drift ice field and the length scale of ice stress. At a compact ice edge, there can be a temperature/salinity front in the boundary layer below the ice and another temperature front in the atmospheric boundary layer.

The mechanical properties of drift ice depend primarily on its compactness and thickness [1, 40, 41]. Open water and thin ice provide more freedom for motion and deformation, while compact, thick ice gives strong resistance to deformation. For uniform circular floes, dense packings are between $\frac{\pi}{4} \approx 0.79$ and $\frac{\pi}{2\sqrt{3}} \approx 0.91$ depending on the arrangements of floes, but denser packing is possible for a distributed floe size. Further compression necessitates ice breakage and pressure ice formation. Sea ice floes break continuously into smaller pieces and freeze together in the cold season to form larger ones. For the continuum approach, the particle size should be one order of magnitude larger than the floe size.

8.4.2 Sea Ice Thickness Distribution

The spatial distribution of ice thickness, $\xi = \xi(x, y; t)$, including the fraction of open water with $\xi = 0$, represents the material state of drift ice [68]. Denoting by $S(\xi)$ the area of ice thinner than

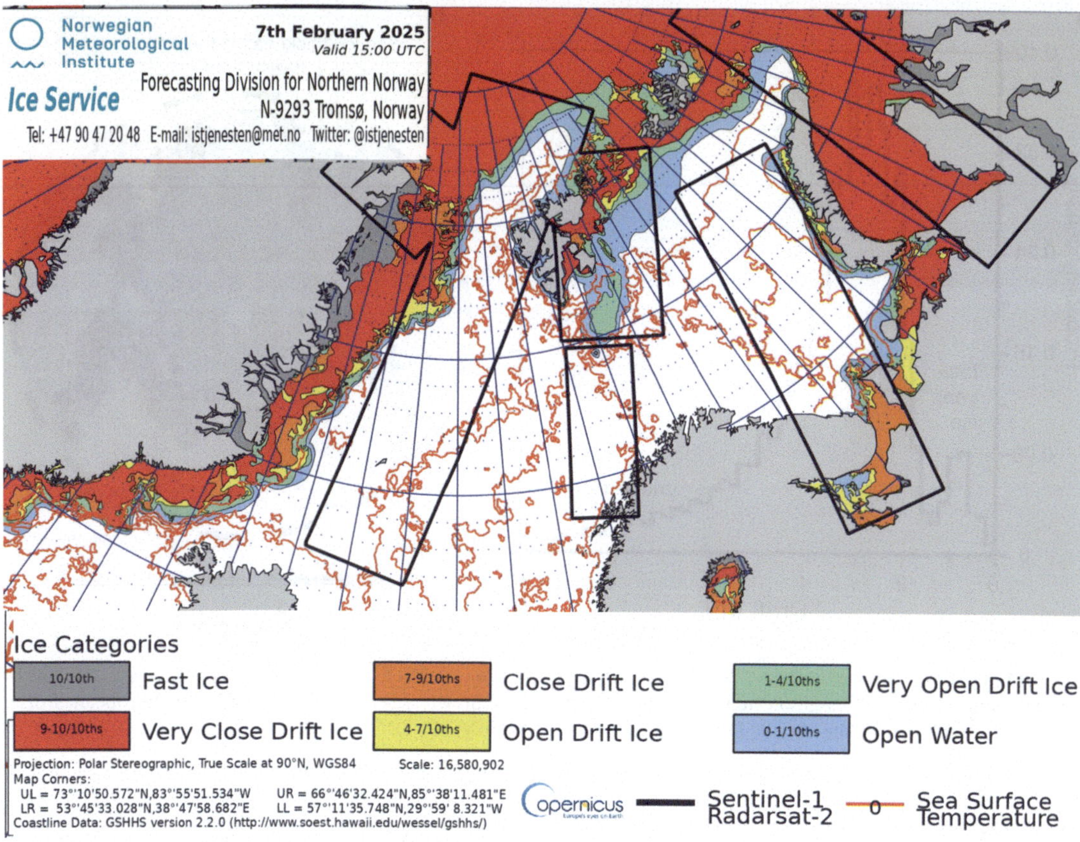

Fig. 8.10 Ice chart over North Atlantic on February 7, 2025. (*Source* Operational sea ice service of the Norwegian Meteorological Institute)

or equal to ξ, the cumulative thickness distribution is

$$\Pi(\xi) = \frac{S(\xi)}{S(\infty)} \tag{8.8}$$

Thus, $\Pi(0)$ is the open water fraction, and $\Pi(\infty) = 1$. The distribution can be written as a sum of its discrete and continuous parts as

$$\Pi(\xi) = \sum_{\xi_k \le \xi} \pi_k + \int_0^{\xi} \pi'(\xi')d\xi' \tag{8.9}$$

where $\pi_k' s$ are the discrete steps corresponding to patches of constant thickness $\xi = \xi_k$, and π' is the continuous part of the distribution. E.g., $\xi_0 = 0$, and $\pi_0 = 1 - A$. The distribution function has no uniform general form, so discrete histograms are used. Figure 8.11 shows ice thickness distribution from the Arctic Ocean. The multi-year peak is well-developed and originates from the Beaufort Sea.

Ridges form the tail of the thickness distribution. Their spatial distribution is described in terms of their size and occurrence. The mean ridge spacing ranges from 100 to 200 m in the Arctic Ocean [69]. Assuming spatial isotropy, the areal mean thickness of ridges is obtained from [70]

$$h_r = \frac{\pi}{2}\mu c H_r^2 \tag{8.10}$$

where μ is the ridge density equal to the inverse mean spacing, H_r is the mean ridge size (keel depth + sail height), and c~2 is a coefficient depending on the geometry of ridges. In the shear zone, ridge orientation is anisotropic because the coastline's geometry strongly

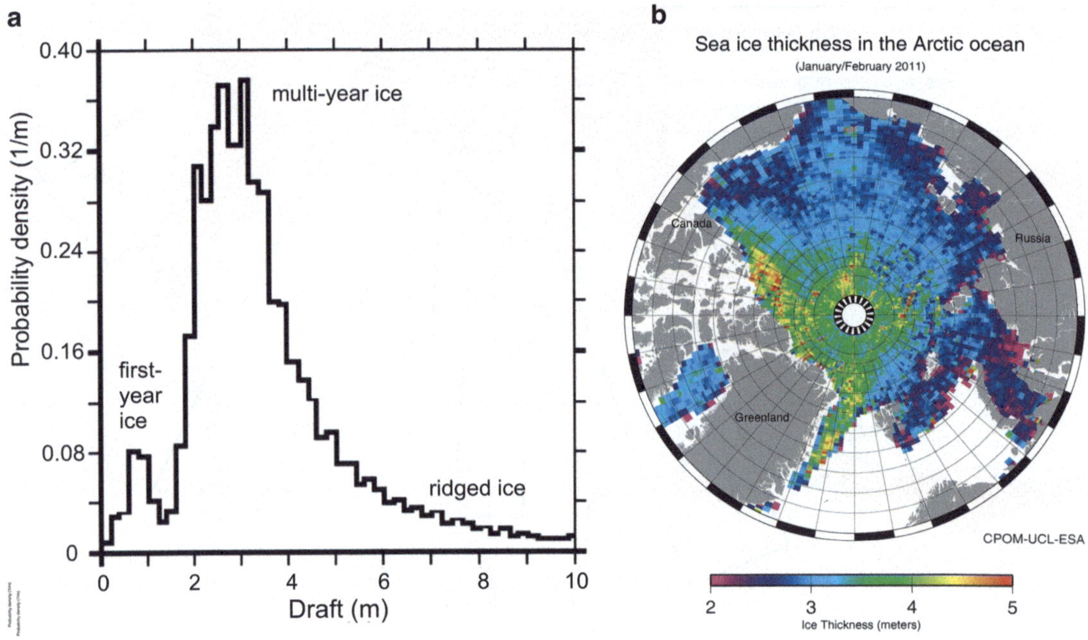

Fig. 8.11 (**a**) Observed ice draft (≈ 0.9 × thickness) distribution in the Fram Strait based on submarine sonar data; and (**b**) ice thickness in the Arctic Ocean in January–February 2011 based on Cryosat data. (*Sources* **a** Redrawn from Wadhams [52], **b** European Space Agency)

influences the fracturing of ice cover, but Eq. (8.10) gives a good approximation when the ridge density is averaged over different directions. Ridging is the main sink of kinetic energy in ice deformation [71], and ridges are essential hydrodynamic form drag elements at the air–ice and ice–water interfaces.

8.4.3 Ice Conservation Law

An ice conservation law can be formulated for the thickness distribution. The density of ice is taken as a constant in drift ice studies. For illustrative purposes, taking a two-class distribution with ice compactness and thickness $\{A, H\}$ [72, 73], we have:

$$\frac{\partial H}{\partial t} + \boldsymbol{u} \cdot \nabla H = -H \nabla \cdot \boldsymbol{u} + \phi_H \qquad (8.11\text{a})$$

$$\frac{\partial A}{\partial t} + \boldsymbol{u} \cdot \nabla A = -A \nabla \cdot \boldsymbol{u} + \phi_A \quad (0 \leq A \leq 1) \qquad (8.11\text{b})$$

where ϕ_H and ϕ_A are the thermal growth/melt rates of ice thickness and compactness, respectively. The left-hand sides contain local change and advection, and the right-hand sides contain mechanical deformation and thermal change. A general formulation has been prepared with the spatial density of ice thickness, $\pi = \pi(\xi)$:

$$\frac{\partial \pi}{\partial t} + \boldsymbol{u} \cdot \nabla \pi = \psi - \pi \nabla \cdot \boldsymbol{u} + \frac{\partial}{\partial \xi}(\phi_\xi \pi) \qquad (8.11\text{c})$$

where ψ is the thickness redistributor and ϕ_ξ is the growth rate of thickness ξ [68]. Compared with (8.11a, 8.11b), the redistributor is an extra term transferring ice from thin classes to thick classes in compression. This is made by assuming thickness multiplication by a factor of 5–15.

8.5 Sea Ice Dynamics

8.5.1 Theory

The drift of a single ice floe depends on its size and shape, but the resulting velocity variance between floes is small due to compensating effects between the roughness of the top and bottom surfaces. When ice floes close, they group and lose much of their individuality, and internal stress becomes significant. As a field of floes, drift ice is generally considered a continuum that limits the spatial resolution to larger than the characteristic floe size [1, 40]. Apart from the resolution, the floe size and shape are not crucial since a system of floes shows significant internal variance only at low concentrations.

A 'point' in a drift ice field is a finite cell where the ice is described by its thickness Π, and lateral velocity $\boldsymbol{u}$ (see [1]). Winds and ocean currents drive sea ice, while the ice responds to forcing through its inertia, internal friction, and the adjustment of its thickness field. The general form of the equation of motion is (Fig. 8.12)

$$\rho_i H\left(\frac{\partial \boldsymbol{u}}{\partial t} + \boldsymbol{u}\cdot\nabla\boldsymbol{u} + f\boldsymbol{k}\times\boldsymbol{u}\right) = \nabla\cdot\boldsymbol{\sigma} + \boldsymbol{\tau}_a + \boldsymbol{\tau}_w - \rho_i H g \nabla\zeta \tag{8.12}$$

where f is the Coriolis parameter, $\boldsymbol{k}$ is the unit vector vertically upward, $\boldsymbol{\sigma}$ is the 2D ice stress, $\boldsymbol{\tau}_a$

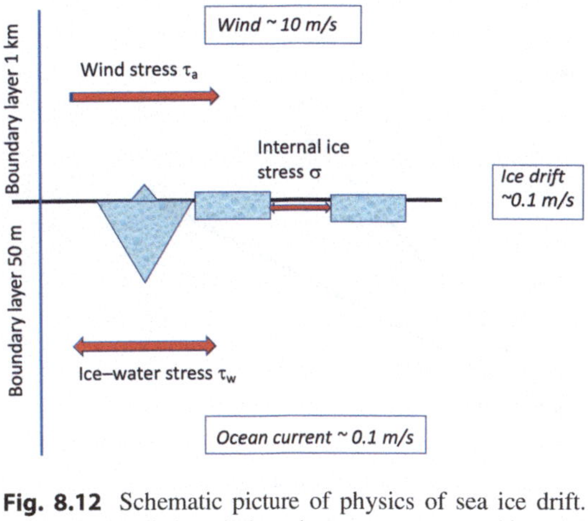

Fig. 8.12 Schematic picture of physics of sea ice drift. The ice is driven by winds and ocean currents, and internal friction arises from the ice stress field

and $\boldsymbol{\tau}_w$ are the air and water stresses on the top and bottom surfaces of ice, respectively, and ζ is the sea surface elevation (e.g., [1, 40]). The Coriolis parameter is $f = 2\Omega \sin\phi$, where Ω is the angular velocity of the Earth, and ϕ is the latitude, $\phi > 0$ ($\phi < 0$) in the northern (southern) hemisphere. The lateral boundary of drift ice can be solid, where the ice velocity is zero, or open, where the ice does not support normal stresses. The open boundary condition is often removed, taking open water as 'ice with zero thickness'. Landfast ice is usually taken as the solid boundary. However, modelling the decay of landfast ice can be seriously biased when static condition is assumed (see [49]).

The air–ice and water–ice stresses are given (e.g., [45, 74, 75]) by the quadratic drag laws

$$\boldsymbol{\tau}_a = \rho_a C_a u_a (\cos\theta_a + \sin\theta_a \boldsymbol{k}\times)\boldsymbol{u}_a \tag{8.13a}$$

$$\boldsymbol{\tau}_w = \rho_w C_w |\boldsymbol{u}_w - \boldsymbol{u}|(\cos\theta_w + \sin\theta_w \boldsymbol{k}\times)(\boldsymbol{u}_w - \boldsymbol{u}) \tag{8.13b}$$

where C_a and C_w are the air and water drag coefficients, θ_a and θ_w are the boundary layer angles in air and water, and $\boldsymbol{u}_a$ and $\boldsymbol{u}_w$ are the wind and current velocities. Representative values of the parameters are for the geostrophic flow, in neutral stratification, $C_a = 1.5 \cdot 10^{-3}$, $C_w = 5.5 \cdot 10^{-3}$, and $\theta_a \approx \theta_a \approx 25°$ [74, 75]. In general, the drag parameters depend on the surface roughness and the stability of the stratification of the boundary layer.

Table 8.4 provides a magnitude analysis of the terms of the equation of motion (Eq. 8.12) based on the typical scales for time $\mathcal{T}$, length $\mathcal{L}$, ice thickness $\mathcal{H}$, and ice velocity $\mathcal{U}$ [1]. The key term is the internal friction. It is important at high ice compactness but insignificant when the compactness is low. The latter case is called *free drift*, for which the solution is easy to find, but the presence of internal friction results in a nonlinear system of partial differential equations. Wind stress is the main driving force, largely compensated by the ice–water stress and the internal friction, while the Coriolis effect becomes important for thick ice. There are three principal time scales: local acceleration $\mathcal{T}_I = \mathcal{H}/(C_w\mathcal{U})$,

Table 8.4 Scaling of the equation of motion of drift ice. The representative elementary scales are ice thickness $\mathcal{H} = 1$ m, ice velocity $\mathcal{U} = 10$ cm s^{-1}, ice strength $\mathcal{P} = 50$ kPa, wind velocity $U_a = 10$ m s^{-1}, sea surface slope $\nabla\zeta = 10^{-6}$, time $\mathcal{T} = 1$ day, and horizontal length $\mathcal{L} = 100$ km

Term	Scale	Value mPa	Comments
Local acceleration	$\rho_i \mathcal{H}\mathcal{U}\mathcal{T}^{-1}$	1	25 for rapid changes ($\mathcal{T} = 1$ h)
Advective acceleration	$\rho_i \mathcal{H}\mathcal{U}^2\mathcal{L}^{-1}$	0.1	Usually much less than water stress
Coriolis term	$\rho_i \mathcal{H} f \mathcal{U}$	10	50 for $\mathcal{H} = 3$ m and $\mathcal{U} = 20$ cm s^{-1}
Internal friction	$\mathcal{P}\mathcal{H}\mathcal{L}^{-1}$	0–500	Negligible in free drift
Air stress	$\rho_a C_a U_a^2$	250	Mostly significant
Water stress	$\rho_w C_w \mathcal{U}^2$	50	200 for $\mathcal{U} = 20$ cm s^{-1}
Pressure gradient	$\rho_i \mathcal{H} g \nabla \xi$	5	Mostly less than Coriolis term

advection $\mathcal{T}_D = \mathcal{L}/\mathcal{U}$, and inertial period f^{-1}, which are well separated: $\mathcal{T}_I \ll f^{-1} \ll \mathcal{T}_D$. Shallow water waves, the Coriolis effect, and tides cause periodic variations. Inertia is small for time scales more than a few hours, and advective acceleration is mostly much smaller than the ice–water stress.

8.5.2 Free Drift

Free drift is a good approximation at ice compactness less than about 80%. Then, internal friction is negligible and momentum advection is very small (Table 8.4), and we have the first-order differential equation,

$$\rho_i H\left(\frac{\partial \boldsymbol{u}}{\partial t} + f\boldsymbol{k}\times\boldsymbol{u}\right) = \boldsymbol{\tau}_a + \boldsymbol{\tau}_w - \rho_i H g \nabla\zeta \tag{8.14}$$

It is advantageous to employ the complex variable presentation,[1] with ice velocity given as $u = u_x + iu_y$, etc., where $i = \sqrt{-1}$ is the imaginary unit. Taking the geostrophic flows as the references in the drag formula (8.13a and 8.13b) and expressing the sea surface tilt as $ifu_{wg} = -g\nabla\zeta$, the free drift equation is written (e.g., [1])

$$\rho_i H\left[\frac{du}{dt} + if\left(u - u_{wg}\right)\right] = \rho_a C_a e^{i\theta_a}|u_a|u_a - \rho_w C_w e^{i\theta_w}\left|u - u_{wg}\right|\left(u - u_{wg}\right) \tag{8.15}$$

Geostrophic current represents the flow independent of the frictional influence of ice. Since the inertial time scale is short, the steady state solution is a very good approximation:

$$u = ae^{-i\theta}u_a + u_{wg} \tag{8.16}$$

The first term on the right-hand side is the wind-driven ice velocity, where a is the wind factor, and θ is the deviation angle between the directions of wind-driven ice drift and wind (Fig. 8.13). Nansen [13] and Ekman [14] presented the solution of wind-driven ice drift based on the drift of Nansen's *Fram* ship in the

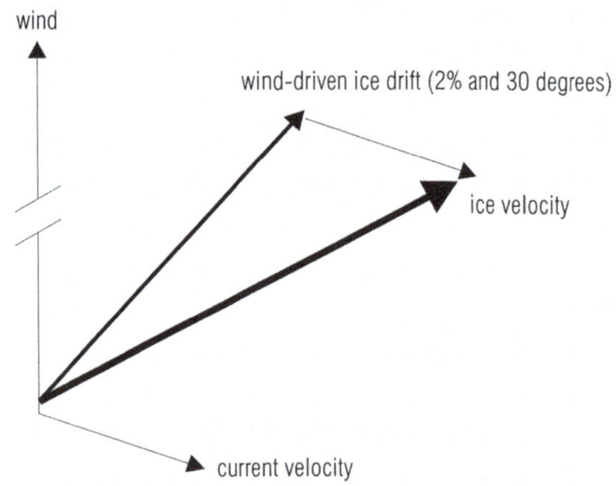

Fig. 8.13 The free drift solution as a vector sum of wind-driven drift and geostrophic ocean current

[1] Rotation by angle ϕ is made using the Euler's formula $e^{i\phi} = \sin\phi + i\cos\phi$.

Arctic and Ekman's boundary layer theory. They obtained $a_s = 2\%$ and $\theta_s = 30°$ for the surface wind, which correspond to the geostrophic values $a = 1.2\%$ and $\theta = 5°$. When ice is thin, the governing balance is between the drag forces and asymptotically $a \to \sqrt{\frac{\rho_a C_a}{\rho_w C_w}} \approx 1.7\%$, $\theta \to \theta_w - \theta_a \approx 0$ for $h \to 0$.

8.5.3 Internal Friction of Drift Ice

The free drift model cannot be combined with an ice conservation law for a full basin sea ice model. Mechanical ice accumulation is not limited in free drift, and an unrealistic ice thickness field would result. Ice stress needs to be included in the system to solve this question. This stress is due to interaction (piling, friction) between ice floes and is written in general form $\boldsymbol{\sigma} = \boldsymbol{\sigma}(\Pi, \boldsymbol{\varepsilon}, \dot{\boldsymbol{\varepsilon}})$, where $\boldsymbol{\varepsilon}$ is strain and $\dot{\boldsymbol{\varepsilon}}$ is strain rate. The stress divergence is an internal force in the dynamics of the system (Eq. 8.12). When sea ice concentration is less than about 0.8, internal stress is small, as shown by the floe collision modelling of Shen et al. [76]. Granular models have a direct physical basis to derive the stress, but they have difficulties on large scales, and all present sea ice climate models are based on the continuum approximation.

Example 8.2 In linear 3D incompressible fluids, such as liquid water, the internal stress $\boldsymbol{\sigma}$ consists of hydrostatic pressure p and viscous stress, $\boldsymbol{\sigma} = -p\boldsymbol{I} + 2\mu\dot{\boldsymbol{\varepsilon}}'$, where $\boldsymbol{I}$ is the unit tensor, μ is the viscosity, and $\dot{\boldsymbol{\varepsilon}}' = \dot{\boldsymbol{\varepsilon}} - \frac{1}{3}(tr\dot{\boldsymbol{\varepsilon}})\boldsymbol{I}$ is the deviatoric strain rate. The internal force is from this stress equal to $\nabla \cdot \boldsymbol{\sigma} = -\boldsymbol{\nabla} p + \mu\nabla^2\boldsymbol{u}$. The first term on the right-hand side is the pressure gradient, and the second is viscous internal friction.

A realistic rheology of compact drift ice ($A > 0.8$) has the following general properties: (1) Ice strength is sensitive to ice compactness and thickness; (2) Yield strength $\gg 0$; (3) Compressive strength > shear strength; (4) Tensile strength is small; and (5) No memory. Drift ice rheology has a distinct asymmetry in that the stress may be very high in closing but nearly zero in opening. Plastic laws were introduced in the 1970s in the AIDJEX (Arctic Ice Dynamics Joint Experiment) programme, with yield strength increasing with ice thickness [73, 77]. A computationally effective, widely applied version is the elastic-viscous-plastic model of Hunke and Dukewicz [78]. Another approach is an elastic-brittle medium proposed by Girard et al. [79], assuming that ice deformation is based on multiscale fracturing and friction, giving a more realistic structure for ice deformation vs. length scale [80].

The elastic-brittle rheology allows a more detailed and physically based description of the fracturing of ice cover. At the same time, classical plastic modelling provides a feasible solution to sea ice dynamics with few parameters. Viscous-plastic models are well applicable for long-term simulations, and they have three key physical parameters: compressive strength, shear strength, and a compactness coefficient. The strength of drift ice depends on the thickness distribution, with the thinnest ice breaking first [77]. Drift ice is strain hardening in compression. Therefore, pressure ice formation may proceed only to a specific limit. A practical form based on the compactness and mean thickness, generalized from Hibler [73], is

$$P = P_n H^n \exp[-C(1 - A)] \tag{8.17}$$

where P_n and n are the compressive strength parameters and C is the compactness coefficient, which tells how strength is reduced when opening. The power n varies in the range $1/2 \leq n \leq 2$, depending on the mode of deformation and $C \gg 1$ due to the high sensitivity of strength to compactness. In the original version, $n = 1$ and $C = 20$.

Two-dimensional plastic stress is specified with a yield curve $F(\sigma_1, \sigma_2) = 0$, where σ_1 and σ_2 are the principal stresses (Fig. 8.14). Drucker's postulate (see [77]) for stable materials states that the yield curve serves as the plastic potential, and according to the associated flow rule, the failure strain is directed perpendicular to the yield curve (e.g., [82]). When the ice fails, the plastic flow is obtained from the equation of motion. To solve this in numerical modelling, stresses below the yield level (inside the yield curve, see Fig. 8.14)

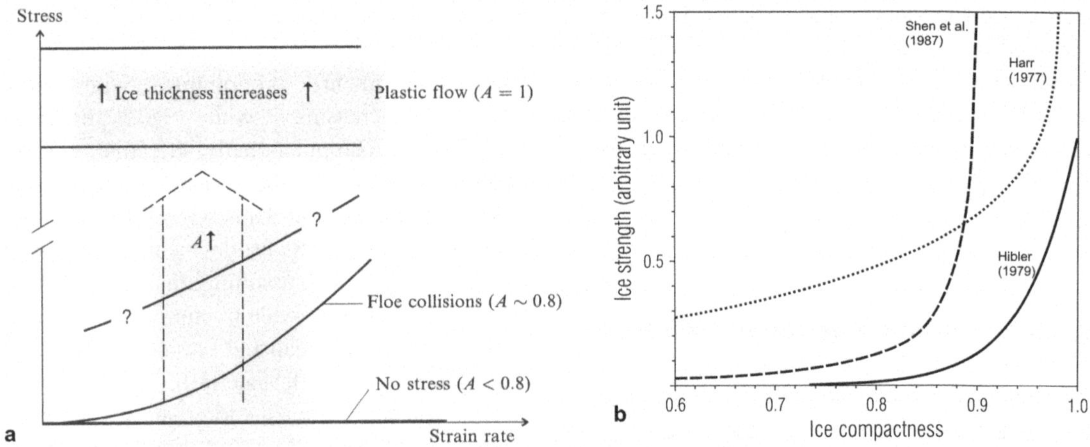

Fig. 8.14 Rheology of drift ice. (**a**) Schematic presentation of change in the quality of sea ice rheology as a function of ice compactness A and thickness. The cut in the ordinate axis corresponds to a jump of several orders of magnitude. (**b**) Compressive strength of drift ice as a function of compactness for floe collision model [76], floe contact model Harr [81], and standard version of Hibler [73] sea ice rheology

must be modelled for a functional relation of stress vs. strain. In a pure plastic model, the stress is indeterminate at zero strain. An elastic sub-model was used in the AIDJEX model [77], a viscous sub-model in Hibler [73], and an elastic–viscous sub-model in Hunke and Dukewicz [78].

Example 8.3 Consider a channel with a uniform ice cover with thickness H from $x = 0$ to a closed boundary $x = L$. Forced by wind $\tau_a > 0$, the dynamics of a plastic model is given by

$$\rho_i H\left(\frac{\partial u}{\partial t} + u\frac{\partial u}{\partial x}\right) = \frac{\partial \sigma}{\partial x} - \rho_w C_w u^2 + \tau_a$$

$$\frac{\partial H}{\partial t} + u\frac{\partial H}{\partial x} = -H\frac{\partial u}{\partial x}$$

The ice breaks and drifts if $\tau_a L > \sigma_Y(H)$ or the yield stress, and the ice pack is shifted towards the positive direction. With $\sigma(0) = 0$, the integral of the quasi-steady momentum equation gives $u = \sqrt{\frac{1}{\rho_w C_w}\left(\tau_a - \frac{\sigma_Y}{L}\right)}$, and ice build-up begins at the solid boundary. The situation develops stepwise in a quasi-steady model, σ_Y increasing with H until equilibrium with $u \equiv 0$. The final length of the ice cover is $\int_d^L H_\infty dx = HL$ where $H_\infty = H_\infty(x)$, $d \le x \le L$, represents the equilibrium thickness. Thus, the ice edge has shifted by the distance d, and the volume Hd has been used to build-up the ice pack. The geometry of the build-up zone depends on the functional form of $\sigma_Y = \sigma_Y(H)$.

8.5.4 Sea Ice Models

Sea ice is a particular air–sea interface. In high latitudes, the momentum, heat, and matter exchanges between the atmosphere and the ocean go through sea ice fields, which undergo dynamic and thermodynamic processes. This is crucially important to regional weather and climate analysis and forecasting. In the practical world, four major questions are connected to sea ice models: short-term forecasting for tactical navigation [20, 83], ice forcing on ships and fixed structures [37], pollutant transport by drifting sea ice [84], and oil spills and oil combating in ice-covered seas [85, 86].

The first realistic sea ice climate model was presented by Hibler [73]. The mean annual velocity from the simulation and the mean thickness field for the Arctic Ocean corresponded well when validated against submarine data (Fig. 8.15). The model predicted that there is

Fig. 8.15 The average annual ice velocity (top) and April thickness contours (bottom) produced by Hibler's [73] model. (Published by the American Meteorological Society [73], printed with permission)

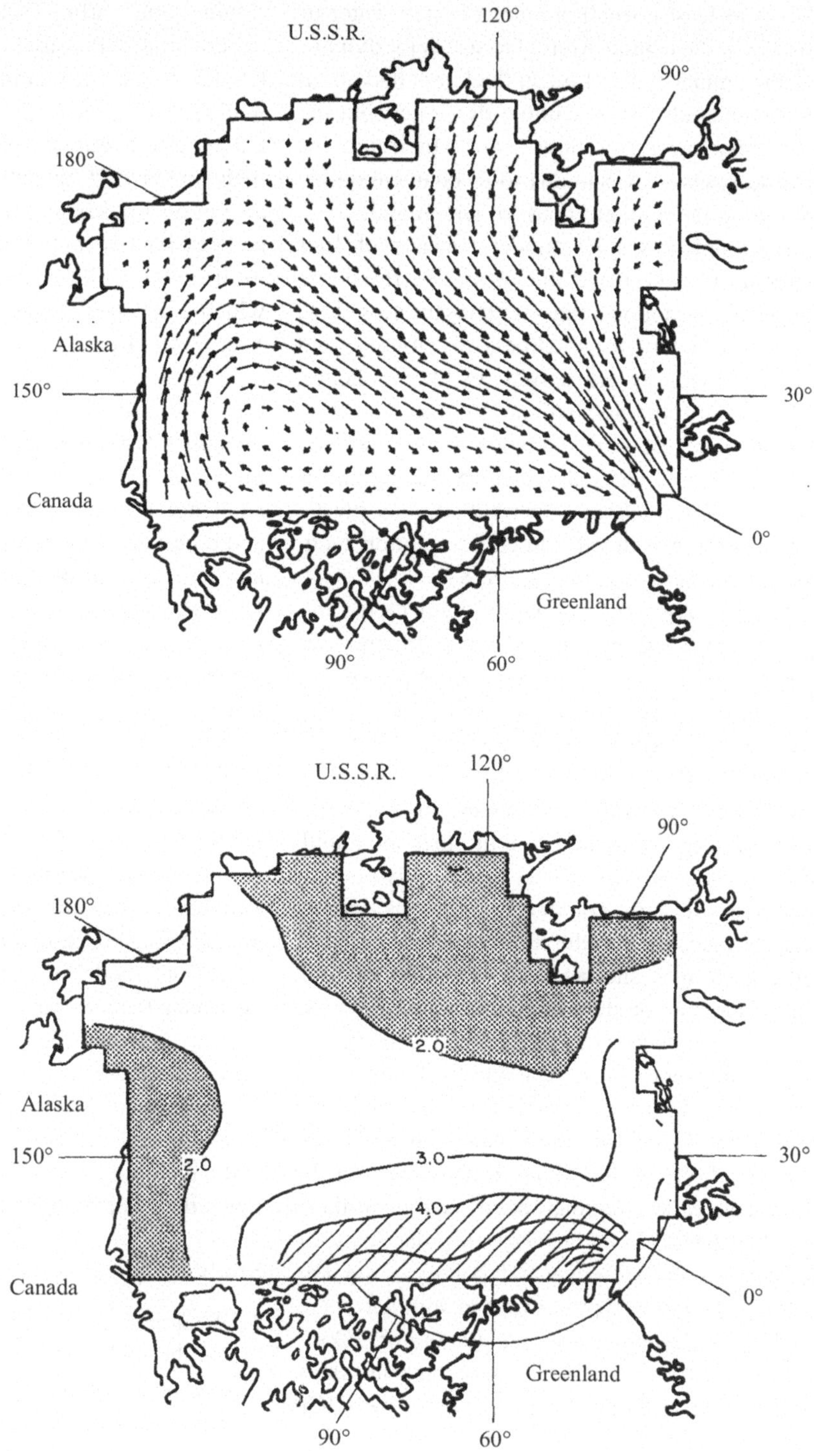

always at least a small amount of open water or thin ice in the central Arctic Ocean. With dynamics, the annual cycle of mean thickness had about the same amplitude as without dynamics, but the level was lower by about 1 m. Maximum ice thicknesses were 5–8 m due to mechanical deformation on the northern coast of Greenland.

A complete sea ice model consists of four elements (see [1]): ice thickness, ice rheology, equation of motion, and ice conservation law. Categorisation of ice thickness distribution and drift ice rheology are chosen by the modeller, depending on the application, and the conservation laws of ice and momentum are fixed by nature. Usually, a multi-category histogram is used for the thickness distribution and plastic rheology with a tuned numerical solution technique are employed [73, 77, 78, 87].

Sea ice dynamics is coupled with thermodynamics since ice growth and melt influence ice strength while the differential motion of ice influences further growth and melt. The primary drift velocity's geophysical parameters are the ice strength and the air–ice and water–ice drag coefficients. The drag coefficients tune the free drift velocity, while the strength tunes the length scale of ice motion as well as the mobility of compact ice. The ice conservation law is largely straightforward, but the thickness redistributor suffers from a shortage of validation data in multi-category models. In climate research, the extent of ice plays a key role in the albedo effect. Sea ice transports latent heat and fresh water, and its melting forms a considerable heat sink and freshwater source into the oceanic surface layer. The variability of this source is a concern for the deep convection regions, such as the northern North Atlantic. In the ecology of polar seas, the location of the ice edge and polynyas has key roles in the primary production in summer.

The Los Alamos National Laboratory Community Ice CodE (CICE) model represents state-of-the-art sea-ice modelling (e.g., [88, 89]). CICE uses the elastic-viscous-plastic rheology [78], advanced atmospheric and oceanic drag parameterisations [90], an incremental linear remapping for the ice advection term, and computes the ice thickness redistribution through ridging and rafting schemes by assuming an exponential redistribution function. In version CICE5, the sea ice salinity profile is prognostic, while snow on the ice is assumed to be fresh [91]. CICE5 also features explicit parameterizations for melt ponds [88], anisotropic rheology [92], and an option for sea-ice biogeochemical modelling and passive tracers [93]. CICE provides an option to use the Delta-Eddington radiation scheme, which calculates apparent optical properties (see Sect. 2.4.2) from inherent optical properties [94].

Complex climate models incorporated our current understanding of sea-ice-related climate dynamics [95]. A realistic simulation of the evolution of sea ice in a sea-ice–ocean model is challenging because sea ice is a highly sensitive and relatively thin layer, and disturbances in oceanic energy and freshwater fluxes can result in large deviations in the distribution of sea ice [96, 97]. To provide increasingly realistic simulations of marginal and seasonal ice zones, floe size distributions have been included to consider ice-wave interactions, such as the breakup of floes due to the ocean wave action and the attenuation of waves when progressing into an ice field [2, 98].

Louvain-La-Neuve (LIM) [99] and CICE are two sea-ice models widely used for large-scale and mesoscale sea-ice modelling. They are used in climate modelling frameworks, coupled with an ocean model and often with an atmospheric model, and they are part of operational sea-ice nowcasting and forecasting systems [100]. Their success is mainly due to the inclusive community development and freely distributed open-source code, including documentation and test cases that make their use particularly easy.

Study Problems

1. Assume that for an ice floe of initial thickness h_0, the melt rate is CuT_w where C is the heat exchange coefficient, u is the ice velocity, and T_w is the water temperature. How far the floe can go if (a) $T_w =$ constant, and (b) $T_w = a \cdot x$. What would be the optimal velocity in terms of travelled distance? Solve the problems in general form and using the parameter

values $h_0 = 1$ m, $C = 10^{-6}$ J m^{-3} °C^{-1}, $u = 10$ cm s^{-1}, and (a) $T_w = 1$ °C, (b) $a = 1$ °C/100 km.

2. Derive the equation of brine volume for ice warmer than −8.2 °C (salt crystals can be neglected):

$$\nu_b = \frac{\rho_i S_i}{\rho_b S_b}$$

where ρ is the density, S is the salinity, and the subscripts i and b are for sea ice and brine, respectively. Assuming that $\rho_i = 920$ kg m^{-3} and $\rho_b = (1000 + 0.82S_b)$ kg m^{-3}, determine the brine volume for $T = -8$ °C & $S_i = 10$‰, C, $T = -5$ °C & $S_i = 5$, 10, 20‰, and $T = -2$ °C & $S_i = 10$‰.

3. How much freezing and melting can influence the salinity and density of a 30-m oceanic mixed layer?
4. An ice floe drifts free in the northern hemisphere. The water velocity is 7 cm/s south, and the west wind blows 15 m/s. Determine the speed and direction of the floe.
5. In a horizontal channel, ice is moving at a steady state speed u driven by the wind stress. When the wind ceases, how far the floe will travel before stopping?
6. Hibler [73] viscous-plastic rheology is written

$$\boldsymbol{\sigma} = \frac{P}{2}\left[\left(\frac{\dot{\varepsilon}_I}{\dot{\varepsilon}_D} - 1\right)\mathbf{I} + \frac{1}{e^2\dot{\varepsilon}_D}(\dot{\boldsymbol{\varepsilon}} - \dot{\varepsilon}_I\mathbf{I})\right],$$

$$\dot{\varepsilon}_D = \sqrt{\dot{\varepsilon}_I^2 + (\dot{\varepsilon}_{II}/e)^2}$$

where P is the pressure, e is the ratio of compressive strength to shear strength, and $\dot{\varepsilon}_I = \dot{\varepsilon}_{xx} + \dot{\varepsilon}_{yy}$ and $\dot{\varepsilon}_{II} = \sqrt{(\mathrm{tr}\dot{\boldsymbol{\varepsilon}})^2 + 4\det\dot{\boldsymbol{\varepsilon}}}$ are strain-rate invariants (see Eq. 2.7). How is the ice stress and velocity in steady-state ice drift along a straight coast? Take y-axis along the coast, then velocity must be a function $u_x = 0$, $u_y = u_y(x)$; why is this necessary?

References

1. Leppäranta M (2011) The drift of sea ice, 2nd edn. Springer, Berlin
2. Wadhams P (2000) Ice in the ocean. Gordon & Breach Science, London
3. Weeks WF (2010) On sea ice. University of Alaska Press, Fairbanks
4. Zubov NN (1945) L'dy Arktiki [Arctic ice]. Izdatel'stvo Glavsermorputi, Moscow. [English translation 1963 by U.S. Naval Oceanographic Office and American Meteorological Society, San Diego]
5. Thomas DN (ed) (2016) Sea ice, 3rd edn. Wiley-Blackwell, Hoboken
6. Pounder E (1965) Physics of ice. Pergamon Press
7. Olaus Magnus Gothus (1555) Historia de gentibus septentrionalibus [A description of the northern peoples], Rome
8. Angnatok J, Laing R (2018) Sea ice. Indigenous peoples atlas of Canada. The Royal Canadian Geographical Society, Ottawa
9. Wadhams P (2016) A fairwell to ice. A report from the Arctic. Oxford University Press, Oxford
10. Weeks WF (1998) The history of sea ice research. In: Leppäranta M (ed) Physics of ice-covered seas, vol 1. Helsinki University Press, Helsinki, pp 1–24
11. Weyprecht K (1879) Die Metamorphosen des Polareises [The metamosphosis of polar ice]. The Austro–Hungarian Polar expedition of 1872–1874. Moritz Perles, Vienna
12. Stefan J (1891) Über die Theorie der Eisbildung, insbesondere über Eisbildung im Polarmeere. Ann Phys 42(2):269–286
13. Nansen F (1902) The oceanography of the North Polar Basin. Norwegian north polar expedition 1893–1896. Scientific results (vol III, no 9). Longman Green & Company, Kristiania
14. Ekman VW (1902) Om jordrotationens inverkan på vindströmmar i hafvet. Nyt Magasin för Naturvidenskab, B 40:1
15. Brennecke W (1921) Die ozeanographischen Arbeiten der Deutschen Antarktischen Expedition 1911–1912. Archiv der Deutschen Seewarte, XXXIX Jahrgang, Nr. 1
16. Gordon AL (1993) Weddell sea exploration from ice station. EOS Trans Am Geophys Union 74(11): 121–126
17. Pritchard RS (ed) (1980) Proceedings of ICSI/AIDJEX symposium on sea ice processes and models. University of Washington, Seattle
18. Untersteiner N (ed) (1986) The geophysics of sea ice. Plenum Press, New York
19. Jurva R (1937) Über die Eisverhältnisse des Baltischen Meeres an den Küsten Finnlands nebst einem Atlas [On the ice conditions in the Baltic Sea at the Finnish coastal area]. Merentutkimuslait.Julk./Havsforskningsinst. Skr. 114. Finnish Institute of Marine Research, Helsinki
20. Leppäranta M (1981) An ice drift model for the Baltic Sea. Tellus 33(6):583–596
21. Fukutomi T (1952) Study of sea ice (the 18th report): drift of sea ice due to wind in the Sea of Okhotsk, especially in its southern part. Low Temp Sci 9:137–144

22. Tabata T (1972) Radar network for drift ice observations in Hokkaido. In: Karlsson T (ed) Sea ice: proceedings of international conference. National Research Council, Reykjavik, pp 67–71
23. Weeks WF, Ackley SF (1986) The growth, structure and properties of sea ice. In: Untersteiner N (ed) Geophysics of sea ice. Plenum Press, New York, pp 9–164
24. Palosuo E (1961) Crystal structure of brackish and fresh-water ice. Publication 54, pp 9–14. IASH Commission of Snow and Ice
25. Weeks WF, Gow AJ, Kosloff P, Digby-Argus S (1990) The internal structure, composition and properties of brackish ice from the Bay of Bothnia. In: CRREL monograph, vol 90–1, pp 5–15
26. Eicken H, Lange M (1989) Development and properties of sea ice in the coastal regime of the southern Weddell Sea. J Geophys Res 94:8193–8206
27. Martin S (1981) Frazil ice in rivers and oceans. Annu Rev Fluid Mech 13(1):379–397
28. Granberg HB (1998) Snow on sea ice. In: Leppäranta M (ed) Physics of ice-covered seas, vol 2. Helsinki University Press, Helsinki, pp 605–649
29. Langhorne PJ, Hughes KG, Gough AJ, Smith IJ, Williams JM, Robinson NJ, Stevens CL, Rack W, Price D, Leonard GH, Mahoney AR, Haas C, Haskell TG (2015) Observed platelet ice distributions in Antarctic sea ice: an index for ocean-ice shelf heat flux. Geophys Res Lett 42:5442–5451
30. Assur A (1958) Composition of sea ice and its tensile strength. In: Thurston W (ed) Arctic sea ice, US National Academy of Science/National Research Council, Washington, DC, pp 106–138. Publ. No. 598
31. Cox GFN, Weeks WF (1983) Equations for determining the gas and brine volumes in sea ice samples. J Glaciol 29(102):306–316
32. Leppäranta M, Manninen T (1988) The brine and gas content of sea ice with attention to low salinities and high temperatures. Internal report 1988(2). Finnish Institute of Marine Research, Helsinki
33. Timco GW, Frederking RMW (1996) A review of sea ice density. Cold Reg Sci Technol 24(1):1–6
34. Horner R, Ackley SF, Dieckmann GS, Gulliksen B, Hoshiai T, Legendre L, Melnikov IA, Reebugh WS, Spindler M, Cornelius CW (1992) Ecology of sea ice biota. Polar Biol 12:417–427
35. WMO (2014) WMO sea-ice nomenclature, terminology, codes and illustrated glossary. WMO/OMM/BMO 259. World Meteorological Organization, Geneva
36. Timco GW, Burden RP (1997) An analysis of the shapes of sea ice ridges. Cold Reg Sci Technol 25: 65–77
37. Palmer A, Croasdale K (2012) Arctic offshore engineering. World Scientific Publishing, Singapore
38. Kovacs A, Sodhi DS (1980) Shore ice pile-up and ride-up: field observations, models, theoretical analyses. Cold Reg Sci Technol 2:210–288
39. Wright B, Hnatiuk J, Kovacs A (1978) Sea ice pressure ridges in the Beaufort sea. In: Proceedings of IAHR ice symposium, Luleå, Sweden
40. Hibler WD III (1986) Ice dynamics. In: Untersteiner N (ed) Geophysics of sea ice. Plenum Press, New York, pp 577–640
41. Rothrock DA (1975b) The mechanical behavior of pack ice. Annu Rev Earth Planet Sci 3:317–342
42. Rees WG (2005) Remote sensing of snow and ice. CRC Press, Taylor & Francis Group, Boca Raton
43. Bitz CM, Lipscomb WH (1999) An energy-conserving thermodynamic model of sea ice. J Geophys Res Oceans 104(7):15669–15677
44. Schwerdtfeger P (1963) The thermal properties of sea ice. J Glaciol 4(36):789–807
45. Curry JA, Webster PJ (1999) Thermodynamics of atmospheres and oceans. Academic Press, New York
46. Makshtas AP (1984) The heat budget of Arctic ice in the winter. Gidrometeoizdat, Leningrad [English transl. by International Glaciological Society, Cambridge, 1991]
47. Vancoppenolle M, Fichefet T, Goosse H, Bouillon S, Madec G, Morales Maqueda MA (2009) Simulating the mass balance and salinity of Arctic and Antarctic sea ice. 1. Model description and validation. Ocean Modell 27:33–53
48. Semtner AJ (1976) A model for the thermodynamic growth of sea ice in numerical investigations of climate. J Phys Oceanogr 6(3):379–389
49. Yang Y, Leppäranta M, Li Z, Cheng B, Zhai M, Demchev D (2015) Model simulations of the annual cycle of the landfast ice thickness in the East Siberian Sea. Adv Polar Sci 26(2):168–178
50. Yang Y, Li Z, Leppäranta M, Cheng B, Shi L, Lei R (2016) Modelling the thickness of landfast sea ice in Prydz Bay, East Antarctica. Antarctic Sci 28(1): 59–70
51. Reimnitz E, Eicken H, Martin T (1995) Multiyear fast ice along the Taymyr Peninsula, Siberia. Arctic 48(4):359–367
52. Wadhams P (1981) The ice cover in the Greenland and Norwegian seas. Rev Geophys 19(3):345–393
53. Leppäranta M (1993) A review of analytical sea ice growth models. Atmos Ocean 31(1):123–138
54. Gow AJ, Ackley SF, Weeks WF, Govoni JW (1982) Physical and structural characteristics of Antarctic sea ice. Ann Glaciol 3:113–117
55. Maykut GA, Untersteiner N (1971) Some results from a time-dependent, thermodynamic model of sea ice. J Geophys Res 76:1550–1575
56. Leppäranta M (1983) A growth model for black ice, snow ice and snow thickness in subarctic basins. Nord Hydrol 14(2):59–70
57. Saloranta T (2000) Modeling the evolution of snow, snow ice and ice in the Baltic Sea. Tellus 52A:93–108
58. Cheng B, Mäkynen M, Similä M, Rontu L, Vihma T (2013) Modelling snow and ice thickness in the

coastal Kara Sea, Russian Arctic. Ann Glaciol 54(62):105–113
59. Shirasawa K, Leppäranta M, Saloranta T, Polomoshnov A, Surkov G, Kawamura T (2005) The thickness of landfast ice in the Sea of Okhotsk. Cold Reg Sci Technol 42:25–40
60. Makshtas AP, Bogorodskiy PV, Kustov VY (2012) Rapid melt of landfast ice in Sogo Bey (Tiksi Gulf) during spring 2011. Probl Arctic Antarctic 1:37–47. (in Russian)
61. Weeks WF (1980) Overview. Cold Reg Sci Technol 2:1–35
62. Goldstein RV, Osipenko NM, Leppäranta M (2009) Relaxation scales and the structure of fractures in the dynamics of sea ice. Cold Reg Sci Technol 58:29–35
63. Leppäranta M (2013) Land-ice interaction in the Baltic Sea. Estonian J Earth Sci 62(1):2–14
64. Volkov VA, Johannessen OM, Borodachev VE, Volnov GN, Petersson LH, Bobylev LP, Kouraev AV (2002) Polar seas oceanography. An integrated study of the Kara Sea. Springer-Praxis, Berlin
65. Hibler WD III, Ackley SF, Crowder WK, McKim HL, Anderson DM (1974) Analysis of shear zone ice deformation in the Beaufort Sea using satellite imagery. In: Reed JC, Sater JE (eds) The coast and shelf of the Beaufort Sea. The Arctic Institute of North America, Arlington, pp 285–296
66. Squire V (1998) The marginal ice zone. In: Leppäranta M (ed) Physics of ice-covered seas, vol 1. Helsinki University Press, Helsinki, pp 381–446
67. Wadhams P (1980) Ice characteristics in the seasonal sea ice zone. Cold Reg Sci Technol 2:37–87
68. Thorndike AS, Rothrock DA, Maykut GA, Colony R (1975) The thickness distribution of sea ice. J Geophys Res 80:4501–4513
69. Hibler WD III, Weeks WF, Mock SJ (1972) Statistical aspects of sea-ice ridge distributions. J Geophys Res 77:5954–5970
70. Mock SJ, Hartwell A, Hibler WD III (1972) Spatial aspects of pressure ridge statistics. J Geophys Res 77: 5945–5953
71. Rothrock DA (1975a) The energetics of the plastic deformation of pack ice by ridging. J Geophys Res 80(33):4514–4519
72. Doronin YP (1970) On a method of calculating the compactness and drift of ice floes. Trudy Arkticheskii i Antarkticheskii Nauchno-issledovatel'skii Institut 291:5–17 [English transl. 1970 in AIDJEX Bull 3: 22–39]
73. Hibler WD III (1979) A dynamic-thermodynamic sea ice model. J Phys Oceanogr 9:815–846
74. Andreas EL (1998) The atmospheric boundary layer over polar marine surfaces. In: Leppäranta M (ed) Physics of ice-covered seas, vol II. Helsinki University Press, Helsinki, pp 715–773
75. McPhee MG (2008) Air–ice–ocean interaction: turbulent ocean boundary layer exchange processes. Springer, Berlin
76. Shen HH, Hibler WD III, Leppäranta M (1986) On applying granular flow theory to a deforming broken ice field. Acta Mech 63:143–160
77. Coon MD, Maykut GA, Pritchard RS, Rothrock DA, Thorndike AS (1974) Modeling the pack ice as an elastic-plastic material. AIDJEX Bull 24:1–105
78. Hunke E, Dukewicz JK (1997) An elastic-viscous-plastic model for sea ice dynamics. J Phys Oceanogr 27:1849–1867
79. Girard L, Bouillon S, Weiss J, Amitrano D, Fichefet T, Legat V (2011) A new modelling framework for sea-ice mechanics based on elasto-brittle rheology. Ann Glaciol 52:123–132
80. Schulson E (2004) Compressive shear faults within arctic sea ice: fracture on scales large and small. J Geophys Res 109:C07016
81. Harr ME (1977) Mechanics of particulate media. A probabilistic approach. McGraw-Hill, New York
82. Davis RO, Selvadurai APS (eds) (2012) Elasticity and geomechanics. Cambridge University Press, Cambridge
83. Hebert DA, Allard RA, Metzger EJ, Posey PG, Preller RH, Wallcraft AJ, Phelps MW, Smedstad OM (2015) Short-term sea ice forecasting: an assessment of ice concentration and ice drift forecasts using the U.S. Navy's Arctic Cap Nowcast/Forecast System. J Geophys Res Oceans 120:8327–8345
84. Pfirman SL, Eicken H, Bauch D, Weeks WF (1995) The potential transport of pollutants by Arctic sea ice. Sci Total Environ 159:129–146
85. Ovsienko S, Zatsepa S, Ivchenko A (1999) Study and modelling of behaviour and spreading of oil in cold water and in ice conditions. In: Proceedings of 15th conference on port and ocean engineering under Arctic conditions, Espoo, Finland
86. Wang K, Leppäranta M, Gästgifvars M, Vainio J, Wang C (2008) The drift and spreading of Runner-4 oil spill and the ice conditions in the Gulf of Finland, winter 2006. Estonian J Earth Sci 57(3):181–191
87. Coon MD (1980) A review of AIDJEX modeling. In: Pritchard RS (ed) Proceedings of ICSI/AIDJEX symposium on sea ice processes and models. University of Washington, Seattle, pp 12–23
88. Flocco D, Schroeder D, Feltham DL, Hunke EC (2012) Impact of melt ponds on Arctic sea ice simulations from 1990 to 2007. J Geophys Res 117 (C9):C09032
89. Holland PR, Kwok R (2012) Wind-driven trends in Antarctic sea-ice drift. Nat Geosci 5:1–4
90. Tsamados M, Feltham DL, Schroeder D, Flocco D (2014) Impact of variable atmospheric and oceanic form drag on simulations of Arctic sea ice. J Phys Oceanogr 44(5):1329–1353
91. Turner AK, Hunke EC (2015) Impacts of a mushy-layer thermodynamic approach in global sea-ice simulations using the CICE sea-ice model. J Geophys Res Oceans 120:1253–1275

92. Tsamados M, Feltham DL, Wilchinsky AV (2013) Impact of a new anisotropic rheology on simulations of Arctic sea ice. J Geophys Res Oceans 118:91–107
93. Jeffery N, Hunke EC, Elliott SM (2011) Modeling the transport of passive tracers in sea ice. J Geophys Res 116(C7):1–15
94. Briegleb BP, Light B, Briegleb BP, Light B (2007) A Delta-Eddington multiple scattering parameterization for solar radiation. In: The sea ice component of the community climate system model, Ncar Tech. Note NCAR/TN-472+STR
95. Griffies SM, Biastoch A, Böning C, Bryan F, Danaba-Soglu G, Chassignet EP, England ME, Gerdes R, Haak H, Hallberg RW, Hazeleger W, Jungclaus J, Large WG, Madec G, Pirani A, Samuels BL, Scheinert M, Sen Gupta A, Severijns CA, Simmons HL, Treguier AM, Winton M, Yeager S, Yin J (2009) Coordinated ocean-ice reference experiments (COREs). Ocean Model 26:1–46
96. Dethloff BK, Abegg C, Rinke A, Hebestadt I, Romanov VF (2001) Sensitivity of Arctic climate simulations to different boundary-layer parameterizations in a regional climate model. Tellus 5:1–26
97. Uotila P, O'Farrell SPO, Marsland SJ, Bi D (2012) A sea-ice sensitivity study with a global ocean-ice model. Ocean Model 1(1):1–59
98. Bennetts LG, O'Farrell S, Uotila P (2017) Brief communication: impacts of ocean-wave-induced breakup of Antarctic sea ice via thermodynamics in a stand-alone version of the CICE sea-ice model. Cryosphere 11:1035–1040
99. Rousset C, Vancoppenolle M, Madec G, Fichefet T, Flavoni S, Barthélemy A, Benshila R, Chanut J, Levy C, Masson S, Vivier F (2015) The Louvain-La-Neuve sea ice model LIM3.6: global and regional capabilities. Geosci Model Dev 8:2991–3005
100. Bertino L, Heimbach P, Blockley E, Ólason E (2024) Numerical models for monitoring and forecasting sea ice: a short description of present status. State of the planet discussions, pp 1–9

Cryosphere and Future 9

Recreation and tourism are increasingly important in snow and ice environments

9.1 This Book

This book has comprehensively introduced the geophysics of snow and ice on Earth or the Earth's cryosphere. Snow is a form of ice but distinguished due to its different qualities and appearances. Snow and ice are dominant factors in the atmosphere, on water surfaces, and in the ground in cold regions of nature. The polar ice sheets in Antarctica and Greenland contain by far most of the ice volume, while seasonal snow and sea ice form the largest visible area of the cryosphere on the Earth's surface. In general, snow and ice occur where the air temperature of

M. Leppäranta, P. Uotila, *Geophysics of the Cryosphere*, Springer Textbooks in Earth Sciences, Geography and Environment, https://doi.org/10.1007/978-3-031-93071-3_9

the coldest month is below the freezing point of pure water, and perennial ice can be found in tundra and polar climates where summers are cold [1–3].

Snow forms in clouds and falls on the Earth's surface, building up the snow cover. Seasonal snow is a temporary water storage in the hydrological cycle and largely modifies the local environment by its high albedo, low light transmittance, and low thermal conductivity [4]. Where the seasonal snow survives over summer, long-term accumulation leads to the formation of glaciers. The thickness of large, continental ice sheets is kilometres, limited by the strength of the ice, and they regulate the global mean sea level and contain large freshwater storages [5]. In cold climates, the water in the ground freezes in winter. It forms the zone of frozen ground, which is seasonal in mid-latitudes but perennial, permafrost in polar and tundra climates, extending to hundreds of metres deep [6].

Floating ice forms in cold environments in lakes, rivers, and seas. According to Archimedes' law, the ratio of freeboard to draft is about 1:10. Floating ice grows statically from the bottom of the ice or dynamically as frazil ice in streams. In inland waters, the ice is almost everywhere seasonal while in high-polar seas multi-year ice exists. The thickness of floating ice is limited to the order of metres by the large latent heat of freezing and low thermal conductivity of ice, and the perennial sea ice growth is additionally limited by drifting of ice to lower latitudes [7]. Lakes and rivers freeze when the winter temperature falls below the freezing point, while sea ice boundaries are limited by the coastline and oceanic polar front [8]. Consequently, sea ice is found above the latitude 60° except in the Northeast Atlantic, where the North Atlantic Current keeps the ocean open to 77°N west of Svalbard.

Northern peoples adapted to live in a snow and ice environment in prehistoric times. In northern Europe and America, after the Last Ice Age, the retarding ice sheet opened good hunting and fishing grounds, and following this process, traces of human life have been found. Skis and sledges were used 9000 years ago in northern Eurasia, much before the wheel was invented (Fig. 9.1), and winter fishing methods were available. Snowshoes were common in northeast Asia and North America for travelling over snow-covered terrain. Inuit people adapted their life in an extreme environment in the presence of both seasonal and perennial ice. During the Little Ice Age, snow and ice were present in winter all over Central Europe as illustrated in many historical paintings. Polar expeditions were very active in the nineteenth century, culminating in reaching the poles in 1909–1912. Parallel to geographical findings, geophysical polar research also grew, as seen in the First International Polar Year in 1882–1883.

Measurements have shown that since the Little Ice Age (1450–1850), the global climate has been warming, and the response of the cryosphere has

Fig. 9.1 Snow and ice have been the base of wide winter communication networks in the north. Sledge has much older history than the wheel. The picture shows a chain of reindeers and sledges in Finnish Lapland in the 1920s. Finnish Heritage Agency, Vapriikki collection. (*Photograph* Gabriel Engberg)

been nearly proportional to the rate of increase of Earth's average near-surface temperature. However, the warming has been regionally stronger in the Arctic than elsewhere due to multiple snow and ice-related feedbacks that amplify the warming [9]. The linear response is evident in the continents where the seasonal cryosphere largely follows the freezing point isotherm, and the perennial cryosphere needs a somewhat colder environment. The oceanic polar fronts also control the extent of sea ice. The details of observed and projected cryospheric changes are discussed below mainly based on the latest IPCC Working Group 1 assessment report [10]. Extended and new observations are presented, allowing for better assessments of past changes and more accurate future projections, which machine learning methods have improved. The Fox-Kemper et al. [11] presentation about the physical changes of the cryosphere is central here.

9.2 Climate and Snow and Ice on Land

Seasonal snow climate is characterized by snow cover extent and duration, snow depth, and snow water equivalent [4]. Terrestrial snow covers 45% of the Northern Hemisphere land area at its winter maximum but only 5% of the Southern Hemisphere; because of this, the Northern Hemisphere snow is focused here. Based on satellite observations, the extent of snow cover has been reduced in Arctic and mountain regions across the Northern Hemisphere since 1978. The trends are negative in all months, with the largest reductions in November, December, March, and May (see Sect. 9.5.3 of Fox-Kemper et al. [11]). The sensitivity of the terrestrial snow cover extent to global average temperature has been estimated to be −1.9 million km^2 $°C^{-1}$. Satellite data combined with ground data records show that the spring snow cover has decreased in the Northern Hemisphere since the 1950s. In contrast, snowfall and net snow accumulation have increased in Antarctica over the twentieth century.

Snow water equivalent (SWE) has declined in the Arctic since 1981, especially in spring. The decline has been evident across North America, while in Eurasia, the trend has been negligible. In general, the SWE trends are consistent with the snow cover extent trends, and there is high confidence that the Northern Hemisphere spring SWE has declined since 1981.

The latest generation CMIP6[1]-class global climate models simulate the snow cover extent realistically and better than the previous versions, except for some outliers (see Sect. 9.5.3 of Fox-Kemper et al. [11]). There is a tendency to have too little snow from November to February. These climate models consistently exhibit a near-linear response of about −8% per °C in snow cover extent to rising mean near-surface air temperature (see Sect. 9.5.3 of Fox-Kemper et al. [11]) that does not depend much on the emission scenario. Based on the model results, the snow retrieval will continue if further climate warming occurs despite a likely increase in snow accumulation in the high Arctic continental regions and the central Arctic Ocean. Due to the yearly time scale of the seasonal snow cover, its recovery occurs as soon as the regional climate cools back.

Accumulating from annual snow gains, the time scale of ice sheets and glaciers is very long. Almost all the world's glaciers have been synchronously retreating since the 1950s, unprecedented in at least the last 2000 years (see Sect. 9.5.1 of Fox-Kemper et al. [11]). In the previous 20 years, the highest regional mass loss rates have been observed in the Southern Andes, New Zealand, Alaska, Central Europe, and Iceland, while the lowest losses have occurred in Mountain Asia, Siberia, and the periphery of Antarctica. Glaciers lost more mass in 2010–2019 than in any other preceding decade.

Mountain and polar glaciers are expected to continue melting for decades to come. They are off equilibrium and will reach a reduced mass balance or disappear even without further

[1] Coupled Model Intercomparison Project (CMIP) of the World Climate Research Programme (WCRP) provides climate projections to understand past, present, and future climate changes. CMIP has become essential to IPCC.

warming. Regions with small glacier coverage, such as Central Europe, Scandinavia, Caucasus, North Asia, Western Canada and USA, and New Zealand, are likely to lose nearly all glaciers but with a minimal contribution to sea level rise. Glaciers in the Antarctic, Arctic Canada, Alaska, Greenland, Svalbard, and the Russian Arctic are expected to experience minor losses, but their sea level rise contribution would be significant.

Ice sheets are climate archives. Ice cores have been used to determine past temperature history and atmospheric constituents based on oxygen and hydrogen isotope ratios and gas bubbles. Moreover, ice cores contain depositions, e.g., from volcanic eruptions. The climate record of ice sheets extends back to 0.1–1 million years, providing invaluable information on past climate variations that can be compared, for example, against the current climate change.

The Greenland Ice Sheet has greatly changed since the Mid-Pliocene Warm Period 3 million years ago when the mean sea level was about 5 m higher than present (see Sect. 9.4 of Fox-Kemper et al. [11]). Paleorecords indicate that the ice sheet has been substantially reduced under the past warmings of 2.5–4 °C. Satellite observations suggest it gained mass until 1980, before its mass balance turned negative. Jakobshavn and Kangerlussiaq outlet glaciers in West Greenland, and more recently Helheim Glacier in East Greenland, show heavy mass losses. The average loss rate was 22 Gt/year in 1980–2012.[2] The ice sheet is losing mass mainly because of ice streams and surface melting, and the loss due to surface mass balance has reached almost 70% of the total. The largest losses have occurred in Northwest and Southeast Greenland. In recent decades, the mass loss has accelerated to unprecedented levels, at least compared to the last 350 years.

The Greenland Ice Sheet is expected to continue losing mass in this century, independent of the amount of greenhouse gas emissions, with the contribution of sea level rise from 0.01 to 0.18 m until 2100. The ice loss is expected to become increasingly dominated by surface melting, because marine margins of the ice sheet retreat, and the ocean-forced dynamic response of the margins will diminish. Scenarios beyond the year 2100 indicate that by 2300, the mass loss would raise the sea level on average by 0.15 m in the low-emission case but up to 1.2 m if emissions continue increasing at the highest possible rate.

In the Mid-Pliocene Warm Period and the Last Interglacial, the West Antarctic Ice Sheet was largely melted (see Sect. 9.4 of Fox-Kemper et al. [11]), and due to the contribution of the Antarctic Ice Sheet, the sea level was about 10 m higher than now. The Antarctic Ice Sheet is losing mass mainly due to the calving of icebergs and basal melting in ice shelves, especially in the West Antarctic outlet glaciers. Moreover, some of these losses have been induced locally by ice shelf disintegration preceded by intense surface melting. The East Antarctic Ice Sheet has also lost mass in the last two decades. The mass loss by basal melting is greater than the mass gain from the increased snow accumulation on the continent. These losses amounted to a 7.4 mm sea level rise in 1992–2020.[3]

The mass loss from glaciers and ice sheets became the dominant contributor to the global mean sea level rise in 2006–2018, surpassing the thermal expansion of the warming ocean. Melting glaciers contributed 22% to the sea level rise in 1971–2018. In this period, Greenland and Antarctica contributed 13% and 7%, respectively, and the numbers have increased to 22% and 14% since 2016.

Climate projections rather consistently predict that the Antarctic Ice Sheet will continue losing mass beyond 2100. Even with sustained global warming of 2–3 °C, the West Antarctic Ice Sheet would be almost entirely and irreversibly lost over multiple millennia. This is convincingly backed up by paleoclimate data from previous warm periods (see Sect. 9.4 of Fox-Kemper et al. [11]). It is clear that the probability of complete loss and the rate of losing mass increase

[2] Since the area of Greenland is about 2.2 million km^2, 22 Gtons of ice correspond to 9 mm ice layer on it.

[3] The area of Antarctic continent is 4% of the ocean area so that 7.4 mm sea level rise corresponds to about 200 mm ice layer over the ice sheet.

with increasing surface temperature. However, a widespread Antarctic ice mass loss by 2100 due to the Marine Ice Shelf Instability (see Chap. 6 of this book) is not expected. However, what will happen depends on the behaviour of individual ice shelves, which is very difficult to predict due to sparse observations and limited knowledge of the key physical processes.

A high sea level rise of 2 m by 2100 is improbable but could be caused by an abrupt widespread onset of marine ice cliff instability and marine ice sheet instability (see Chap. 6 of this book) around Antarctica and by a fast change in the surface mass balance and dynamical ice loss from Greenland. The probabilities of disintegration of Greenland glaciers by ice cliff instability cannot be confidently estimated due to uncertainties in deep processes. Observations hint that it should not be expected in Greenland glaciers, where cliffs do not collapse catastrophically but evolve into short floating extensions. This is also the case in vaster Antarctic glaciers. So far, only the Crane Glacier in the Antarctic Peninsula has retreated due to ice cliff instability style after the collapse of the Larsen B ice shelf. At the same time, in Greenland, the risk of this process is considered substantial only for the Jacobshavn and Helheim glaciers.

Accumulating observational evidence and better ice sheet models reveal unexpectedly high melt rates, predicting substantial mean sea level rise in this century (Fig. 9.2). Ice sheets respond slowly to changes in climate warming, and even with a sudden stop of greenhouse gas emissions, they would continue to shrink towards equilibrium. On the other hand, there are few regions, such as the northern Baltic Sea, where land is still slowly rising after the Last Ice Age, locally compensating for the sea level rise. Moreover, when an ice sheet melts, its gravitational pull weakens

Fig. 9.2 Antarctica is the world's snow and ice continent. The picture shows an iceberg calving in the Queen Maud Land coast, Weddell Sea. (*Photograph* Matti Leppäranta)

on the nearby ocean water. The resulting modification of the Earth's gravity field reduces the sea level rise close to the ice sheet but gives more sea level rise further away. These factors, among many others, make the estimation of local sea level trends, extremes, and impacts difficult.

The long-term evolution of permafrost closely follows climate evolution since a common rule is that permafrost occurs where the annual mean air temperature is lower than −2 °C or so. However, ground temperature and permafrost depend on the snow conditions, not just the air temperature. Permafrost warming and thawing have been widespread in the Arctic since the 1980s and are associated with increased active layer thickness. As a result, the near-surface permafrost volume has decreased, and in many discontinuous and sporadic permafrost regions, complete thaw has occurred. In general, permafrost has been more widely present in colder climates than now, and if the global mean air temperature increases, permafrost losses can be expected to continue.

The oldest existing permafrost originates from several interglacial periods, and samples from old permafrost show growth and degradation matching the glacial variability. Paleo modelling studies are consistent with observations suggesting that permafrost was thicker than today during the Last Glacial Maximum below presently glaciated regions but thinner in the permafrost regions that were glaciated then but no more. Such changes manifest the strong thermal insulation of the ground by glacial ice [11].

There is a large scatter among the CMIP climate models regarding their permafrost extent, although some models are relatively close to the observations (see Sect. 9.5 of Fox-Kemper et al. [11]). Models with a permafrost extent that is too low may have too shallow soil columns and do not consider soil water phase changes. In contrast, models that give large permafrost extent may have insufficient insulation by snow cover. However, probably realistically, these models show strong sensitivity of the permafrost volume to global mean near-surface air temperature (see Sect. 9.5 of Fox-Kemper et al. [11]). The amplitude and timing of permafrost response to global warming remain uncertain. The air temperature, integrating the surface heat fluxes, acts as the primary driver of permafrost thermodynamics, and as the temperature increases, permafrost is expected to shrink.

A substantial amount of organic carbon was stored in the Earth's permafrost when it was formed multiple millennia ago. If all this carbon were released into the atmosphere, the carbon dioxide concentration would increase by about 90 ppm.[4] This carbon is released if permafrost thaws in aerobic conditions or thawed soil is unsaturated with liquid water. However, if the conditions are anaerobic or the thawed soil is saturated, carbon is released as methane, which is over 30 times more potent greenhouse gas than carbon dioxide but has a much shorter lifetime (12 years) in the atmosphere. The role of snow is particularly important in thawing the discontinuous permafrost. A thicker snow cover protects the permafrost from melting better, affecting the amount and rate of carbon release from thawing soil.

There is high confidence that the permafrost extent will retreat and the active layer will continue to deepen, if global warming continues. The estimated volume reduction of the perennially frozen ground in the 3-m top layer is about 25% per 1 °C warming with a delayed response in deep permafrost layers. Decreasing permafrost would have hazardous impacts, including carbon release with positive feedback to global warming. Environmental impacts would concern water resources, ecosystems, and built infrastructure. Thawing of ice-rich permafrost and ice wedges would create thermokarst ponds, increase subsurface drainage, and release water pollutants, such as mercury.

9.3 Ice on Inland and Marine Waters

Lake and river ice edge follow closely the January (July) 0 °C isotherm in the Northern (Southern) Hemisphere [8]. Lake ice adapts to the local

[4] The present CO_2 content is 427 ppm, increased by about 100 ppm in the last 50 years.

climate but shows individual characteristics, and deep, geothermal, and highly saline lakes may stay unfrozen in a cold environment. Lake and river ice are seasonal except for some individuals in high Arctic spots and Antarctic dry valleys. The extent of lake and river ice has retreated with the global climate warming in recent decades [12]. The annual ice period has shortened by 1–2 weeks per 1 °C increase in air temperature. The probability of annual freezing is less than one at the ice climate edge, and currently, about 15,000 lakes freeze intermittently.

Many currently seasonally ice-covered lakes will experience occasional ice-free winters if the climate warms [12]. The ice cover of some high Arctic lakes would shift from perennial to seasonal. Moreover, some of the coldest lakes that used to freeze to the bottom would maintain a liquid water body under ice year-round. For example, a warming of 2 °C would more than double the number of intermittently freezing lakes, and reductions in ice-covered periods have been estimated to be 15–40 days in Canada by 2050. The numerous impacts of these changes in lake ice concern fishing, cultural traditions, transportation, and recreation. Their impacts on the hydrological cycle affect hundreds of millions of people [12].

The duration of river ice seasons has decreased by approximately 6.1 days per 1 °C of atmospheric warming [13]. The largest river ice reductions in the past 40 years have occurred in November and March, they were strongest regionally in Alaska and Central Eurasia [13]. The river ice extent is expected to continue declining proportionally to the projected surface air temperature. High spring flows and potential winter thaws are projected to heighten the threat of ice jams while reducing the viability of ice roads and recreational use of the frozen environment. Reductions in the duration of the thick lake and river ice will change the regulation of carbon emissions from rivers and influence ecosystems and recreational activities. Therefore, they could be vital for traditional indigenous communities.

Sea ice extent in the polar oceans has been monitored regularly by passive microwave satellite imagery since 1978. It has been observed that the Arctic sea ice extent has declined every month, particularly since the 1990s (see Sect. 9.3 of Fox-Kemper et al. [11]). The sea ice area has experienced the largest reduction of 25% from August to October, around its annual minimum. Since 2007, the rate of decline has levelled. Based on long-term reconstructions, the current ice extent has been unprecedentedly low, at least since 1850. Regionally, the Barents Sea ice area has declined most of the winter, while in summer, the sea ice margin has seen a more substantial decline in the Beaufort, Chukchi, East Siberian, and Kara Seas than elsewhere. The albedo feedback has accelerated the summer losses, while in winter, the Barents Sea ice losses are likely related to increased atmospheric and oceanic heat transport from the North Atlantic.

The melting of Arctic sea ice has affected the sea ice cover beyond its extent and concentration. Sea ice has become thinner, a smaller amount survives over summer, and there is less multi-year ice. Thinner ice is more mobile, and, in addition, first-year ice is saltier and weaker than multi-year ice, making it more prone to deformation and ridging. Ice thickness changes are challenging to cover with satellite data because they contain limited information on the vertical extent of sea ice. Sonar observations confidently show that Arctic sea ice has become thinner since 1979, and its volume reduced by about 72% during 1979–2016.

The Arctic sea ice cover is currently at its lowest since at least 1950, and it is predicted to reach a 'practically ice-free' summer minimum at least once before 2050 under all emission scenarios. Practically ice-free is defined as the sea ice area below 1 million km^2, about 1/14 of the area of the Arctic Ocean. There is no tipping point for this loss of summer sea ice; if the greenhouse gas concentrations decreased, more sea ice would survive over summer. As a result, the summer sea ice loss is reversible on decadal time scales. With higher emissions, the practically ice-free summer Arctic is projected to occur more often and become the new normal by the end of this century.

Arctic sea ice decline has followed almost linearly the global mean near-surface air temperature increase and, according to climate models, cumulative anthropogenic greenhouse gas emissions. This decline is superimposed on the natural climate variability, which is unknown and partly originates from the variability of oceanic heat transport and partly from atmospheric processes, such as cyclones entering the Arctic. It has been suggested that more than half of the decline in Arctic sea ice in recent decades is due to the anthropogenic impact, and the remainder is due to internal climate system variability.

In the Antarctic, there is no significant trend in the satellite-observed sea ice extent from 1979 to 2020, both winter and summer, due to regionally opposing trends and large internal variability (see Sect. 9.3 of Fox-Kemper et al. [11]). The regionally opposing trends primarily result from the changing regional winds driving sea ice and the changing air temperature that affects ice growth and melting. Since the maximum extent in 2014, the Antarctic sea ice extent has plummeted and has not recovered, with the new all-time minimum reached in 2017, which was replaced by 2023. The reasons for this decline are somewhat unclear but possibly related to the subsurface Southern Ocean circulation and heat fluxes that have intensified in the past few decades, to the levels reducing the sea ice cover [14], but also atmospheric circulation changes have been suggested as a possible cause.

Regarding future Antarctic sea ice projections, climate model simulations provide low confidence due to their inability to reproduce the past evolution manifested by limited agreement with sparse, reliable observational data. This is likely due to deficiencies in process presentation, particularly at the regional level. Due to this low confidence, the attribution of Antarctic sea ice change is problematic, and the contribution of anthropogenic forcing on the sea ice extent remains unclear.

9.4 Closing Words

Our knowledge of snow and ice increased steadily during the twentieth century, along with our interest in nature and the requirements of the environmental information for the modern technology needs in cold regions. In the Arctic, both the Cold War and oil and gas exploration were driving the research, while in the south, the Antarctic Treaty maintained scientific activities. Towards the end of the century and beyond, the increasing awareness of the anthropogenic climate change highlighted the significance of polar zones in the global climate, and more resources were allocated to expeditions, monitoring systems, and model development.

The cryosphere controls the local and global albedo, limits the surface temperature with the freezing point as the upper boundary, and plays a role in the water cycle. The water storage in the ice sheets controls the sea level elevation on a long timescale. In contrast, seasonal snow and sea ice are renewed annually so that they may cause large year-by-year albedo variations amplified by the positive albedo feedback phenomenon. The boundaries of the snow cover, sea ice, frozen ground, and glaciation move along with climatic variations. In a warming climate, melting permafrost releases greenhouse gases, in particular carbon dioxide and methane, thus providing positive feedback.

Future climate projections predict that climate warming will continue [10]. This warming is attributed to human influence, i.e., the industrial flux of greenhouse gases to the atmosphere, and the warming has been quantified by mathematical modelling. As a result, the temperature, precipitation, and the fraction of liquid precipitation are expected to increase in the polar regions. Multiple climate system tipping points, or points of no return, are connected to the cryosphere, most notably the continental ice sheets of Greenland and Antarctica. Given the near-linear response of the cryosphere to the warming and the increasing probability of crossing the tipping points, the importance of limiting the warming becomes evident.

Ice sheets and glaciers are not now in an equilibrium, and therefore they are expected to lose mass further unless the mean local temperature turns down. In a 2–3 °C warmer world, the Arctic Ocean would be practically free of sea ice in September, and the extent of spring snow cover would decrease by 30% from the present. Alpine glaciers could lose a large fraction of their mass in a century, but the time scale of ice sheets is very long to result in large centennial changes. The permafrost volume in the ground top 3 m would experience a large decrease that would have major consequences to the Arctic environment.

The spatial temperature gradient between the January air temperature contours −5 and 0 °C is of the order of 100 km $°C^{-1}$ in Eurasia and North America (Fig. 3.3a). This gives a scale for the geographical change of cry-sphere due to climate warming. However, the solar radiation, determined by the time and place, must be accounted for in the comparisons outside the polar night. Regarding the annual cycle, the air temperature reaches the freezing point in autumn depending on the rate of atmospheric cooling, which is of the order of 1 °C per 5 days, smaller in marine climate and higher in continental climate.

Moreover, the ocean circulation complicates climate forecasting. For example, the permafrost-covered Alaska (60° to 70°N) is located across the same latitudinal band as Scandinavia and Finland in Northern Europe where the climate is 5 °C warmer largely due to the warm North Atlantic Ocean Current, and farming takes place even north of the Arctic Circle. In understanding of the winter cycle of the cryosphere, the most significant uncertainty for climate forecasting lies in the snow cover (Fig. 9.3) due to the extreme role of snow in the thermal control of the surface.

Fig. 9.3 Snow cover is the key cryosphere climate element, its future characteristics form a major uncertainty factor in cryosphere scenarios. (*Photograph:* Matti Leppäranta)

References

1. Barry R, Gan TY (2011) The global cryosphere: past, present, and future. Cambridge University Press, Cambridge
2. Colbeck S (ed) (1980) Dynamics of snow and ice masses. Academic Press, New York
3. Marshall SJ (2011) The cryosphere. Princeton University Press, Princeton
4. DeWalle D, Rango A (2008) Principles of snow hydrology. Cambridge University Press, Cambridge
5. Cuffey KM, Paterson WSB (2010) The physics of glaciers, 4th edn. Academic Press, New York
6. Davis N (2001) Permafrost: a guide to frozen ground in transition. University of Alaska Press, Fairbanks
7. Weeks WF (2010) On sea ice. University of Alaska Press, Fairbanks
8. Leppäranta M (2023) Freezing of lakes and the evolution of their ice cover, 2nd edn. Springer, Berlin
9. Goosse H, Kay JE, Armour KC, Bodas-Salcedo A, Chepfer H, Docquier D, Jonko A, Kushner PJ, Lecomte O, Massonnet F, Park H-S, Pithan F, Svensson G, Vancoppenolle M (2018) Quantifying climate feedbacks in polar regions. Nat Commun 9: 1919
10. IPCC (2021) Climate change 2021: the physical science basis. Contribution of working group I. In: Masson-Delmotte V, Zhai P, Pirani A, Connors SL, Péan C, Berger S, Caud N, Chen Y, Goldfarb L, Gomis MI, Huang M, Leitzell K, Lonnoy E, Matthews JBR, Maycock TK, Waterfield T, Yelekçi O, Yu R, Zhou B (eds) The sixth assessment report of the intergovernmental panel on climate change. Cambridge University Press, Cambridge
11. Fox-Kemper B, Hewitt HT, Xiao C, Aðalgeirsdóttir G, Drijfhout SS, Edwards TL, Golledge NR, Hemer M, Kopp RE, Krinner G, Mix A, Notz D, Nowicki S, Nurhati IS, Ruiz L, Sallée J-B, Slangen ABA, Yu Y (2021) Ocean, cryosphere and sea level change. IPCC (2021), pp 1211–1362
12. Sharma S, Blagrave K, Magnuson JJ, O'Reilly CM, Oliver S, Batt RD, Magee MR, Straile D, Weyhenmeyer GA, Winslow L, Woolway RI (2019) Widespread loss of lake ice around the Northern Hemisphere in a warming world. Nat Clim Chang 9:227–231
13. Yang X, Pavelsky TM, Allen GH (2020) The past and future of global river ice. Nature 577:69–73
14. Purich A, Doddridge EW (2023) Record low Antarctic Sea ice coverage indicates a new sea ice state. Commun Earth Environ 4:1–9

Annexure Ice Properties and Useful Formulae

A1: Properties of Ice Ih and Liquid Water

Table A1.1 Basic properties of freshwater ice as a function of temperature [1]

Quantity	Unit	Temperature 0 °C	−10 °C	−20 °C	−30 °C
Density	kg m^{-3}	916.7	918.7	920.3	921.6
Thermal expansion coefficient, volume	10^{-6} °C^{-1}	159	155	149	143
Adiabatic compressibility	10^{-5} MPa^{-1}	13.0	12.8	12.7	12.5
Specific heat	kJ kg^{-1} °C^{-1}	2.11	2.03	1.96	1.88
Thermal conductivity	W m^{-1} °C^{-1}	2.14	2.3	2.4	2.5
Relative permittivity (dielectric constant)	1	91.6	94.4	97.5	99.7
Latent heat of melting	kJ kg^{-1}	333.5	–	–	–
Latent heat of sublimation	MJ kg^{-1}	2.829	2.832	2.836	2.840

Table A1.2 Basic properties of pure water [2]

Quantity	Unit	Value
Molecular weight	g mol^{-1}	18.016
Gas constant for vapour	J °C^{-1} kg^{-1}	461.5
Temperature of maximum density	°C	3.98
Boiling point	°C	100
Triple point	°C, Pa	0.1, 611.73

Table A1.3 Dependence of freezing point of water on pressure [1]

Quantity	Unit	Value			
Pressure	bar	1	10	100	1000
Melting point	°C	0.00	−0.06	−0.74	−8.80

Table A1.4 Properties of cold liquid water [2]

Quantity	Unit	Temperature 0°C	10°C
Viscosity	N s m^{-1}	$1.793 \cdot 10^{-3}$	$1.307 \cdot 10^{-3}$
Surface tension	N m^{-1}	75.64	74.23
Specific heat of water	kJ kg^{-1} °C^{-1}	4.2176	4.1921
Thermal conductivity	W m^{-1} °C^{-1}	0.561	0.580
Relative permittivity	1	87.90	83.96
Latent heat of evaporation	MJ kg^{-1}	2.494	2.472

Equation of State of Sea Water

The density standard of seawater is TEOS-10 (Thermodynamic Equation of Seawater) based on the absolute salinity and conservative temperature[1] (see [3]). TEOS-10 algorithm cannot be presented in explicit form, but there is free software available in the web (https://www.teos-10.org/). The previous standard, Unesco [4] is an explicit formulation based on temperature (T), practical salinity (S), and pressure (p) to determine the water density

[1] Conservative temperature is used in TEOS-10 for the temperature. It is equal to potential enthalpy divided by fixed heat capacity.

M. Leppäranta, P. Uotila, *Geophysics of the Cryosphere*, Springer Textbooks in Earth Sciences, Geography and Environment, https://doi.org/10.1007/978-3-031-93071-3

ρ_w. This form helps in understanding the density of seawater in general and its dependence on temperature, salinity and pressure:

$$\rho_w(T,S,p) = \frac{\rho_w(T,0,0) + \Delta(T,S,0)}{1 - p/K(T,S,p)}$$

$$\rho_w(T,0,0) = 999.842594 + 6.793952 \cdot 10^{-2}T - 9.095290 \cdot 10^{-3}T^2 + 1.001685 \cdot 10^{-4}T^3 - 1.120083 \cdot 10^{-6}T^4 + 6.536332 \cdot 10^{-9}T^5$$

$$\Delta(T,S,0) = S\Big(0.824493 - 4.0899 \cdot 10^{-3}T + 7.6438 \cdot 10^{-5}T^2 - 8.2467 \cdot 10^{-7}T^3 + 5.3875 \cdot 10^{-9}T^4\Big) + S^{3/2}\Big(-5.72466 \cdot 10^{-3} + 1.0227 \cdot 10^{-4}T - 1.6546 \cdot 10^{-6}T^2\Big) + 4.8314 \cdot 10^{-4}S^2$$

$$K(T,0,0) = 19652.21 + 148.206T - 2.327105T^2 + 1.360477 \cdot 10^{-2}T^3 - 5.155288 \cdot 10^{-5}T^4$$

$$K(T,S,0) = K(0,0,0) + S\Big(54.6746 - 0.603459T + 1.09987 \cdot 10^{-2}T^2 - 6.1670 \cdot 10^{-5}T^3\Big) + S^{3/2}\Big(7.944 \cdot 10^{-2} + 1.6483 \cdot 10^{-2}T - 5.3009 \cdot 10^{-4}T^2\Big)$$

$$K(T,S,p) = K(T,S,0) + p\big(3.239908 + 1.43713 \cdot 10^{-3}T + 1.16092 \cdot 10^{-4}T^2 - 5.77905 \cdot 10^{-7}T^3\big) + pS\big(2.2838 \cdot 10^{-3} - 1.0981 \cdot 10^{-5}T - 1.6078 \cdot 10^{-6}T^2\big) + 1.91075 \cdot 10^{-4}pS^{3/2} + p^2\big(8.50935 \cdot 10^{-5} - 6.12293 \cdot 10^{-6}T + 5.2787 \cdot 10^{-8}T^2\big) + p^2S\big(-9.9348 \cdot 10^{-7} + 2.0816 \cdot 10^{-8}T + 9.1697 \cdot 10^{-10}T^2\big)$$

Values for checking are $\rho_w(0,5,0) = 999.96675$ kg m^{-3}, $\rho_w(35,5,0) = 1027.67547$ kg m^{-3}, and $\rho_w(0,5,0) = 1062.53817$ kg m^{-3}.

Temperature of freezing point T_f [4] and maximum density T_m [5]

$$T_f = -0.0575 \cdot S + 1.710523 \cdot 10^{-3}S^{3/2} - 2.154996 \cdot 10^{-4}S^2 - 7.53 \cdot 10^{-3}p$$

$$T_m = 3.982 - 0.2229 \cdot S - 0.02004 \cdot p(1 + 0.00376 \cdot S)(1 + 0.000402 \cdot p)$$

where $0 \leq S \leq 40‰$, and p is the gauge pressure in bars.

A2: Physical Information of the Environment

Table A2.1 Earth [6]

Quantity	Unit	Value	Comments
Equatorial radius Polar radius	km km	6378.140 6356.755	Mean radius is 6371
Surface area	m^2	5.101·10^{14}	
Mean density	kg m^{-3}	5515	
Mass	kg	5.9722·10^{24}	
Acceleration due to gravity	m s^{-2}	9.80665	9.78 at Equator, 9.83 at poles
Period of rotation	h/min/s	23 h 56 min 4.10 s	
Rotation rate	s^{-1}	0.7292·10^{-4}	
Inclination of ecliptic	deg	23°27′	Also tilt of the rotation axis

Table A2.2 Sun

Quantity	Unit	Value	Comments
Astronomical unit (AU)	m	$149.60 \cdot 10^9$	Mean Sun–Earth distance
Solar constant	$W\ m^{-2}$	1367	
PAR (400–700 nm) fraction	1	0.48	Range ≈ 0.44–0.50

Formulae for the solar year

Sun–Earth distance r $\left(\frac{AU}{r}\right)^2 = 1.000110 + 0.034221 \cos j + 0.001280 \sin j + 0.00719 \cos 2j + 0.000077 \sin 2j;\ j = (J-1)\frac{2\pi}{365}$ rad

Solar altitude angle $h_\odot$ $\sin h_\odot = \sin \phi \sin \delta + \cos \phi \cos \delta \cos \tau$

Declination δ $\sin \delta = \sin \varepsilon \sin\left[(J-80)\frac{2\pi}{365}\right]$

Tilt of Earth's axis $\varepsilon = 23°27'$

Solar hour angle $\tau = \tau_{GMT} - \frac{\Lambda}{15°} + \Delta t,\ \tau = 0$ at local solar noon

Day of year $J = 1.0$ on Jan 1, 00.00 h

Here, ϕ is the latitude, Λ is the longitude, GMT is the Greenwich Mean Time, and Δt is the time correction, $|\Delta t| < 30$ min, which arises from the ellipticy of Earth's orbit around the Sun and is available in astronomical tables.

Table A2.3 Atmosphere [6]

Quantity	Unit	Value
Average molecular mass	$kg\ kmol^{-1}$	28.965
Standard atmospheric pressure	mbar	1013.25
Gas constant of air	$J\ kg^{-1}\ °C^{-1}$	287.04
Kinematic viscosity (0 °C)	$m^2\ s^{-1}$	$13.28 \cdot 10^{-6}$
Thermal diffusivity (0 °C)	$m^2\ s^{-1}$	$19.06 \cdot 10^{-6}$
Specific heat at constant pressure	$kJ\ kg^{-1}\ °C^{-1}$	1.004
Specific heat constant volume	$kJ\ kg^{-1}\ °C^{-1}$	0.717
Thermal conductivity (0 °C)	$W\ m^{-1}\ °C^{-1}$	$2.4 \cdot 10^{-2}$
Saturation water vapour pressure over liquid water	mbar (T in °C)	$\log_{10} e_{sw}(T) = \frac{0.7859 + 0.03477T}{1 + 0.00412T}$
Saturation water vapour pressure over ice	mbar ($T \le 0$ °C)	$\log_{10} e_{si}(T) = \log_{10} e_{sw}(T) + 0.00422T$
Specific humidity	1	$q = \frac{0.622e}{p_a}$; p_a is the air pressure

A3: General Mathematics and Physics

Table A3.1 Angle

Mathematical (right-hand system)	0 toward east, counterclockwise rotation positive, full circle 360° = 2π rad
Geographic	0 toward north, clockwise rotation positive, full circle 360° Mathematical angle = 90° – geographic angle
Wind direction (WD)	Tells where wind blows from in geographic angle. Vector direction = WD + 180° (geographic) or 270° – WD (mathematical)

Table A3.2 Physical constants [2]

Quantity	Unit	Value
Gravitational constant	$N\ m^2\ kg^{-2}$	$6.67 \cdot 10^{-11}$
Avogadro's number	1	$6.022142 \cdot 10^{23}$
Planck's constant	J s	$6.626068 \cdot 10^{-34}$
Boltzmann's constant	$J\ K^{-1}$	$1.3807 \cdot 10^{-23}$
Stefan-Boltzmann constant	$W\ m^{-2}\ K^{-4}$	$5.670400 \cdot 10^{-8}$
Velocity of light in vacuum	$m\ s^{-1}$	$2.99792458 \cdot 10^{8}$
Universal gas constant	$J\ K^{-1}\ mol^{-1}$	8.3145

References

1. Petrenko VF, Whitworth RW (1999) Physics of ice. Oxford University Press, Oxford
2. Line DR (ed) (2001–2002) CRC handbook of chemistry and physics, 82nd edn. CRC Press, Boca Raton
3. Millero FJ, Feistel R, Wright DG, McDougall TJ (2008) The composition of standard seawater and the definition of the reference-composition salinity scale. Deep Sea Res 55(1):50–72
4. UNESCO (1981) The practical salinity scale 1978 and the international equation of state of seawater 1980. Technical papers in Marine Science, 36
5. Caldwell DC (1978) The maximum density points of pure and saline water. Deep Sea Res 25(2):175–181
6. Gill A (1982) Atmosphere–ocean dynamics. Academic Press, New York

M. Leppäranta, P. Uotila, *Geophysics of the Cryosphere*, Springer Textbooks in Earth Sciences, Geography and Environment, https://doi.org/10.1007/978-3-031-93071-3

The manufacturer's authorised representative in the EU is Springer Nature Customer Service Centre GmbH, Europaplatz 3, 69115 Heidelberg, Germany. If you have any concerns regarding our products, please contact ProductSafety@springernature.com

Printed and bound by CPI Group (UK) Ltd, Croydon, CR0 4YY
30/09/2026
02225577-0020